Do Your Own
PLUMBING

Popular Science

Do Your Own PLUMBING

K.E. Armpriester

Published by
Popular Science Books
New York, New York

Published by

Popular Science Books
Grolier Book Clubs, Inc.
380 Madison Avenue
New York, NY 10017

Consultants

Dick Montgomery has over forty years' experience in the
plumbing field. Currently a plumbing inspector for the city of
Dayton, Ohio, Mr. Montgomery is also an instructor of ABC
plumbing (Association of Builders and Contractors, Inc.).

Hans Grigo is the manager of Home Safety for the National
Safety Council, Chicago, Illinois. He is also the NSC house and
yard expert. The National Safety Council is a public service organ-
ization whose mission is to educate and influence society to adopt
safety and health policies, practices and procedures.

Martin M. Mintz, AIA is the Director of Technical Services for
the National Association of Home Builders, Washington, D.C. He
is a registered architect and member of the American Institute of
Architects. Mr. Mintz is a lecturer and writer, and serves on several committees including the NAHB Research Committee and a
National Institute of Building Sciences committee.

Book design by Linda Watts.

Produced by Bookworks, Inc., West Milton, Ohio. Illustrations by
Linda Ball, O'Neil & Associates, and James Schrier.

ISBN:1-55654-031-0

Manufactured in the United States of America.

Acknowledgments

Our thanks to the following companies for the use of materials and supplies:

Air City Plumbing, Dayton, Ohio

Fogle's Plumbing and Heating, West Milton, Ohio

PK Building Center, Englewood, Ohio

Pickrel Bros. Inc., Dayton, Ohio

Rambo-Westendorf Plumbing, Dayton, Ohio

Wertz Hardware, West Milton, Ohio

And to the following companies and individuals for the use of photographic locations: Karen Callahan; Fogle's Plumbing and Heating, West Milton, Ohio; Montgomery County Joint Vocational School, Clayton, Ohio; Pickrel Bros. Inc., Dayton Ohio; Supply Dayton, Dayton, Ohio.

Introduction

This Popular Science book is not merely a reference manual. It has been written and designed especially for you, the home-owner, as a basic how-to manual—to show you how to perform essential home skills—every step of the way.

Do Your Own PLUMBING will instruct and guide you in all the basic plumbing skills that you should know to keep your plumbing modernized and running smoothly. You will learn how to perform basic plumbing skills such as replacements, cutting and connecting pipe, and, most importantly, you'll learn to deal with plumbing problems as they arise.

In the chapter *How Your Plumbing Works,* you'll gain an understanding of your plumbing system. This is an important chapter so please don't neglect to read it. Succeeding chapters will tell you which tools and materials are needed and will teach you the beginning skills that will give you increasing self-confidence. Home plumbing really isn't all that difficult, as you'll soon see.

You'll learn how to do common repairs which will easily save you the amount of money you paid for this book. We have attempted throughout, by the way, to show large, easy-to-interpret color-coded drawings. And we've also conducted research to pro-vide you with diagrams of the most popular and the most practical devices and materials being used in the country today. All this information has been thoroughly reviewed and edited by a plumbing expert and instructor.

With this book, you will also be shown step-by-step how to install new plumbing fixtures in your home—in the bathroom and the kitchen, as well as in utilitarian areas. You'll realize that you can make many plumbing installations just as skillfully and smoothly as a paid plumber.

There are special chapters about faucets and valves and about preparations for plumbing runs. Additionally, we've included a chapter on outdoor piping to show you the special techniques that this kind of plumbing entails.

Home plumbing is a very easy-to-learn skill. With the proper attention to detail, virtually anyone can become their own home plumber. Simply and inexpensively you'll be able to update and improve your home's plumbing system. And, ultimately too, you'll have the irreplaceable joy of knowing that you did it by yourself. Best wishes from Popular Science in this new venture!

John W. Sill
Publisher

Contents

1

How Your Plumbing Works

Put an everyday soda straw into a glass of water, allow it to fill, and then place your thumb over the top. Lift it out of the glass...water will remain in the straw. This simple test demonstrates what might be the most complex principle in home plumbing—pneumatic pressure. Likewise, if you didn't have vent pipes in your home, wastewater could remain in your drainpipes. In this chapter, you'll learn about the principles of plumbing and the three basic plumbing systems—supply, drain-waste, and vent.

Fresh water is a chief concern for everyone these days. Although contaminated water is infrequent in the United States, it is becoming a cause for concern. A troubleshooting guide is provided to help you diagnose and correct any trouble you might have with your water.

Water pressure is rarely a problem. But if you add fixtures you could simultaneously give yourself low water pressure, a condition that's difficult to correct. High-pressure problems, on the other hand, can be fixed by making installations. Plumbing noises are also discussed in this chapter and some remedies are given.

The chapter closes with an explanation of plumbing codes and, most importantly, safety concerns. Electricity combined with water can cause unsafe situations—and these are explained in depth. Also, be sure to read and follow faithfully the first rule of plumbing: *Always shut off the water before you begin your work.* With these warnings, plumbing might strike you as a dangerous activity—but don't be intimidated. It is, in fact, one of the safest home skills that you can acquire.

Understanding the System

Since plumbing, by definition, refers to any work with pipes, in its strictest sense it involves work with gas pipes or heating and cooling pipes. In this book, however, only water system pipes and related gas pipes are addressed. Though it is called the 'plumbing system,' in fact, three distinct systems can be identified by the functions that they perform.

■ The *supply system* carries potable (drinkable) water into your house from an underground city water main, a well, or a spring. It operates under pressure to send water to all the places where you need it. Also in this system are the resources for altering the water to make it more useful—such as heating it or purifying it.

■ The *drain-waste system* takes away the used water from your home and delivers it to a public sewer system or your own septic or seepage tank. Unlike the supply system, it operates by the force of gravity (the few exceptions being with pressure-valve toilets, sump pumps, and sewage ejectors).

■ The *vent system* disposes of sewer gases and also helps to maintain atmospheric pressure within the drainpipes. The vent system is, in fact, connected to the drain-waste system and the two together are sometimes referred to as the *DWV system.*

Once you learn a few basic principles about these three systems, you will begin to understand how they relate to each other and how they connect to the various water-using fixtures and appliances in your home.

How It Comes In—The Supply System. Water from a utility company usually comes from an underground water main, passes through a *water meter,* and then a *stop valve* (also called the *main shutoff valve,* page 9). As its name implies, this valve controls all

UNDERSTANDING YOUR PLUMBING SYSTEM

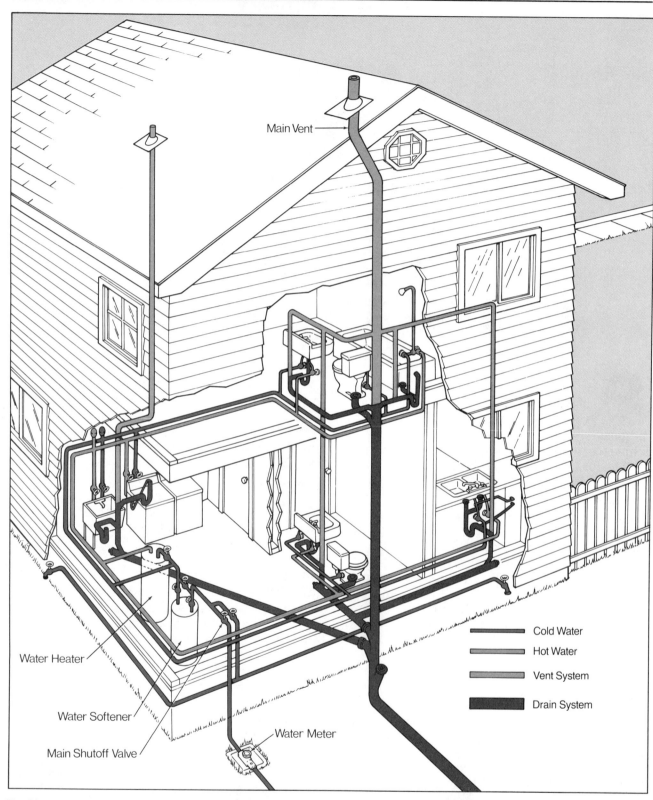

Main Vent

Cold Water

Hot Water

Vent System

Drain System

Water Heater

Water Softener

Main Shutoff Valve

Water Meter

Plumbing systems are all basically the same in that they contain four different kinds of pipes. Use the key as a guide to the four kinds and then follow the routes of the piping. Cold water which is pressurized, enters the home through the water meter (if you have one) and the main shutoff valve. If you have a water softener, cold water may first be routed to outdoor spigots and toilets; then

it will travel on to the softener. Next, cold softened water is sent to fixtures and to the water heater. Hot softened water is also supplied to the home plumbing fixtures—the sinks, tubs and clothes washer. Exiting from all fixtures are the drain-waste pipes which carry, with the help of gravity, all used water and waste out of the house. Most drain-waste pipes are connected to the main soil stack

except ones that connect at the foundation to the main house drain. At the top of the main soil stack begins the main vent stack, a piping system necessary for ridding the water system of gases and fumes, and releasing them into the atmosphere via house rooftops. Every fixture with wastewater must have a venting pipe. Often, as shown here, there is a second, supplementary venting system.

the water entering your home; it should be turned off while working on any supply line that doesn't have a shutoff valve of its own.

IMPORTANT

In some water systems there is an accessible valve located on the 'street side' of the water meter. Called the *meter valve,* this is the water company's responsibility and should never be tampered with. Shutting it off could damage the water meter and jeopardize your dealings with your supplier.

In areas subjected to cold weather, the meter and shutoff valve are located in basements or crawlspaces. In more moderate climates, they can be found in a small concrete well near a curb (called a *dry well*). In some cases the shutoff valve will be located at the property line, such as in a system supplied by a water company but unmetered. If water comes from a private well, the shutoff valve might be located at the wellhead, or at the point where water enters the house.

A typical supply system begins with 1-inch-diameter pipe; cold water enters the home through this pipe under about 50 pounds of pressure per square inch *(psi)*. Close to this point of entry but beyond the meter, are the fixtures that alter the condition of the water. A water filter or a water softener may be installed here. Some people install water softeners at this point to cure hard-water problems in their entire system; others create a bypass in the line for outdoor use, toilets, or a single 'unsoftened' faucet. In any case, the system is divided into hot and cold lines, the diameter of the pipes is gradually decreased, and water is sent to all fixtures and water-using appliances within the home.

The water lines that run vertically, from floor to floor or from the crawl space to the first floor, are called *risers.* Lines running horizontally are called *branches.* Occasionally, at fixtures in a supply line, air chambers or shock absorbers will be added to remedy problems with excessive air pressure (page 7).

How It Goes Out—The Drain-Waste System. Wastewater and sewage are carried off through this system by the force of gravity; however, the layout design of piping is critical to ensure a properly functioning system. Beginning at a fixture, drainpipe is slanted downward toward the *main soil stack,* the primary disposal pipe in a home system. If the slope of

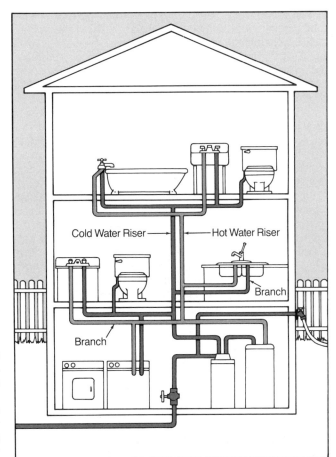

Detail of the Supply System. The supply system, the water coming into the house, is shown here for clarification. Cold, unsoftened water enters the home, goes through the water softener, and branches to fixtures and to the water heater. From the basement, hot and cold softened water runs its course upward via pipes called *risers.*

Cold Water Riser — ← Hot Water Riser

Branch

Branch

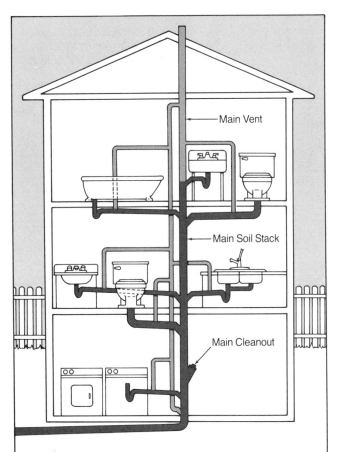

Detail of the Drain-Waste and Vent Systems. The drain-waste and vent systems, sometimes called the *DWV,* are shown here for clarification. Drainpipes for all fixtures are slanted downward to empty into the main soil stack. From there, wastewater and sewage proceeds to the sewer or septic tank. The main cleanout, accessible to the homeowner, is usually located just before this exit point. The piping above the drain-waste system serves as vent piping and it rises above the house in a main or secondary vent stack.

— Main Vent

— Main Soil Stack

Main Cleanout

this drainpipe is too steep, water will run off quickly but solid particles will be left behind. If the pipe is not slanted enough, waste will remain in the system and possibly be backed up into the fixture. The standard pitch for all horizontal drainpipes, therefore, is ¼ inch for every foot of pipe.

Toilets have integral traps; that is, they are drained directly into the main soil stack. Other fixtures have smaller-diameter drainpipe and also use traps as part of the drainage system (page 77). Wastewater from all of the drains flows into the vertical main soil stack which then bends into the slanting, horizontal *main drain*. At this connection there is usually a *main cleanout* in the form of a Y fitting to provide access for ridding the system of large clogs. The main drain is usually located beneath the basement floor; from it wastewater and sewage are carried to a public sewer or a septic tank.

Another Vital Element—The Vent System. In order to maintain atmospheric pressure at every point in

the plumbing system, the free passage of air is essential. Without this air passage a full or partial vacuum can be created causing a 'glug-glug' sound at a drain or, in the worst case, a backup of wastewater at another fixture in the system. This free passage of air is achieved through the presence of vent pipes which, additionally, serve to rid the system of undesirable or noxious sewage gases.

The vent system consists of a *main vent stack* which rises above the main soil stack and passes outdoors through the roof. Smaller-diameter venting pipes, required for each fixture in the system, are extended to this main stack. In larger homes or in single-story homes where it is impractical to run vent pipes great distances to the main vent stack, a group of fixtures may be vented through a *secondary vent stack*.

Traps, though part of the drainage system, also serve in the venting process. They are designed to maintain a constant water level which prohibits

gases from entering the home at the fixtures.

Water Impurities

It would be wonderful to know that by simply turning on a tap you were releasing clean, fresh, and pure water. Unfortunately, this is very rare. Depending on your locale, the groundwater might contain organic or mineral impurities or even contaminants. If water is supplied by a municipal water company, most of these problems should be resolved; however, some, like hard water, will not be. If your source is a personal well, then you must take extra precautions and install any needed purification systems on your own.

Determining the Problem. Most problems in water systems, such as mineral deposits, can be detected visually. Sometimes the sense of taste or smell will trigger a suspicion that something is wrong. Use the troubleshooting chart below to diagnose your problem and determine the solution; or, if you're

TROUBLESHOOTING WATER IMPURITIES

TROUBLE	CAUSE	SOLUTION
White scaly deposits in shower heads and faucets. Soap sludge left in tubs and sinks after use.	Calcium and magnesium compounds. A common problem occurring in approximately 80 percent of the water in the United States.	Install an ion-exchange water softener.
Rust deposits left on clothes after washing. Rusty deposits in sinks and other fixtures. Standing water turns reddish.	Iron compounds. Occurs anywhere; most common in Florida or the Midwest.	For large amounts, install an oxidizing water filter. For small amounts, install an ion-exchange water softener.
Unpleasant 'rotten egg' smell. Black tinge. Silver tarnishes quickly.	Hydrogen sulfide. Occurs anywhere; most common in the Midwest or Southwest.	For large amounts, install a sand filter and a chemical feeder with a chlorine solution. For small amounts, install an oxidizing filter.
Cloudy or dirty water.	Silt, mud, or sand particles. Most common in water drawn from rivers or lakes, also from well water in lowlands.	Install a sand filter.
Reddish 'drip stains' below faucets in fixtures with steel supply pipes. Greenish stains in fixtures with copper pipes.	Carbonic acid. Most common in the East and the Pacific Northwest.	For large amounts, install a chemical feeder with an alkaline solution. For small amounts, install a neutralizing filter.
Unpleasant taste. Brown or yellow tinge.	Organic substance such as algae. Most common in water drawn from rivers or lakes, also from well water in lowlands.	For large amounts, install a chemical feeder with alum or chlorine solution and a sand filter. For small amounts, install a charcoal-core filter.
The presence of bacteria, confirmed by laboratory analysis.	Contamination by nearby sewage disposed of improperly. Most common in well water but also occurs in utility-supplied water.	Correct the sewage disposal problem. Install a chemical feeder with a strong chlorine solution and a charcoal-core filter.

not able to pinpoint it, have it checked by a professional testing laboratory. The absolute worst problem, of course, is contamination, and although this isn't prevalent in the United States, it does occur. The Environmental Protection Agency has numerous lists of cases where people have experienced illness due to unsafe water.

Companies that sell home water-purifiers often give free tests. Local boards of health will test your water for bacteria and often Department of Agriculture extensions will check it for organic matter, acids, or minerals. If all else fails, call a commercial testing lab. If you have a well and there is known contamination in your area, you should have the water tested every six months. After you have installed a fixture to solve your problem, it is still wise to have a lab verify that your new water is acceptable.

Water-Treatment Devices. The most well-known device for treating water in the home is the water softener. Water softeners are beneficial for prohibiting the buildup of minerals such as calcium, magnesium, or iron. Essentially, they are tanks that contain an 'ion-exchange resin' which exchanges sodium for the unwanted mineral(s). Two types of these devices are available. One type is periodically replaced by the supplying company and the other is permanently installed and maintained by the homeowner.

Adding a water softener for the entire water supply is beneficial in that it slows down mineral buildup, saves on the cost of soap, and so forth. However, because water softeners involve the process of adding sodium to the system, they adversely affect the drinking water supply—especially for people with known heart or kidney trouble. For this reason many people position their softeners so that they affect only the hot water system.

Another well-known treatment device is the water filter. These relatively simple devices can be installed for an entire system, usually ahead of the water heater, or at individual fixtures (page 112). There are many types available. Most use either sand, marble chips, or charcoal cores; all of them require periodic back-flushing to remove unwanted particles and routine replacement of their core materials.

The most sophisticated water filter, the reverse-osmosis purifier, is installed under a sink to provide drinking or cooking water from a single tap. It includes a thin membrane through which only pure water can pass. Unwanted particles remain on the other side of the membrane and are drained through a waste outlet.

Chemical feeders are most often installed on private well systems. They contain pumps that inject small amounts of chemicals, such as chlorine, into the supply line. Versatile, these appliances can help to cure a variety of problems.

Problems with Water Pressure

Since most valves, fixtures, and appliances in the home are designed to accept water running at a pressure between 50 and 60 psi (pounds per square inch), it only makes sense to maintain it at that level so that repairs will be kept at a minimum. The problem is that water coming into the home can vary from 10 psi to 150 psi. As a homeowner, you can remedy the situation by adding fixtures to your present system and also by accounting for the water pressure when making new installations. But in extreme cases, the only true cure is to have your pressure adjusted by the local water company.

Problems with water pressure within the home are dependent on a variety of factors which all work together to make the pressure either too high or too low.

■ The number of fixtures. Don't assume that you can add three or four fixtures within your system and not affect the water pressure. Pressure fall-off is very likely to occur in this case.

■ The distance from a fixture to the main supply pipe. The further away from the main supply pipe, the lower the pressure.

■ The distance from one fixture to another. The further away from each other, the lower the pressure.

■ The usage of fixtures. The more fixtures that are being used at the same time, the lower the pressure.

■ The size of piping. The larger the pipe, the higher the *volume*. The water pressure actually remains the same but by enlarging pipes more water is distributed to outlets.

Solving Low-Pressure Problems. Generally speaking, low water pressure is more difficult to correct than high pressure. An obvious sign that your pressure is too low is a thin trickle of water at your faucets. If this happens in your home, but not consistently, the problem might be that you are using your appliances when all of your neighbors are using theirs—at a peak time, in a municipal system that isn't equipped to handle such a demand. If you live in an older home, the problem could be scaly or rusty pipes that are slowing the water down. Your location in relationship to the supplying reservoir, especially if you're on a hill, is another possible cause.

In any case, to solve the problem you will want to take the easiest route possible. Attaining higher water pressure is hardly worth the expense of completely replumbing the house. Two more feasible options, however, are to replace the main supply pipe and to flush the pipes of clogging sediments.

Replacing the Supply Pipe. By replacing the main supply pipe and increasing its diameter, you will increase the volume of water coming into your home. This step should be taken before you consider changing any individual supply pipes that lead to fixtures and appliances. (See page 3 for the location of the main supply pipe.) If the present pipe measures ¾ inch, install a 1-inch pipe. If necessary, call the water company and ask them to install a compatible water meter.

Flushing the Supply System. If you suspect that your low pressure problem is due to clogged pipes, use these steps to flush the system.

1 Remove and clean the aerators on all of your faucets (page 72). Do not replace them.

2 Close the gate valve—either the shutoff valve at the water heater (page 110) or the main shutoff valve (page 9).

3 At the point farthest from the valve, open a faucet as far as it will go.

4 Open another faucet, this one closer to the valve. Now plug it with a rag (but don't turn it off).

5 Reopen the gate valve and allow water to run through the farthest faucet for as long as sediment appears (probably for a few minutes).

6 Turn off the faucets, remove the rag, and replace the aerators.

Solving High-Pressure Problems. Some annoying signals that your water pressure is too high are spurting faucets and clanging sounds when appliances are running. The latter problem, called *water hammer,* occurs when the water flow is shut off by a clothes washer's or dishwasher's valve. The connecting supply pipe, because of the pressure buildup, is shoved into a nearby stud or other building component resulting in the banging or 'hammering' sound.

If your water pressure is extremely high, then you should take the following extra precautions to protect your home from flooding or appliance damage. (Even if it *isn't* high, these are good practices to follow.)

■   Turn off appliance shutoff valves when they are not in use. This is especially important for clothes washer and dishwasher valves.

■   When you go on vacation, turn off the main shutoff valve (page 9).

Another noisy problem results in a screeching sound; this is usually associated with toilet problems but it can also occur in a main line, especially in homes that are located close to a water tower or pumping station. Called *cavitation,* it is caused by an 'unevenness' of pressure within a pipe; bubbles are formed that not only create noise but also erode surfaces.

Sometimes high pressure is simply the result of your home's location. Some cities have extra-high pressure in areas where fire prevention is a chief concern.

If you live in a low-lying area of a mountainous or hilly region, depending on where the supply comes from, you will receive water under high pressure. In these cases, you can contact your water company to let them know of your concern; but more than likely, the situation won't be greatly improved.

High pressure at a water heater can be dangerous, even critical. For this reason, most of them come with temperature and pressure-relief valves. If yours does not have one, be sure to make this installation (page 112).

INSTALLING A PRESSURE-REDUCING VALVE

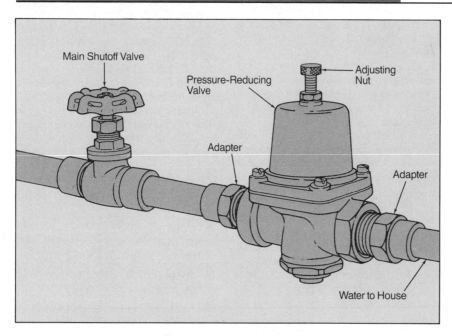

To step down the water pressure in your system install a valve like this in your main water line, near the point where water enters your home, just *after* the main shutoff valve. First, read Chapter 3 to gain the skills for this project. Then, thread adapters of the same material as the pipe into both the inlet and outlet of the valve. Carefully measure the length of the adapter fittings and the valve. (If the line is made of steel, also incorporate the length of unions, page 20.) With the water turned off at the main shutoff valve, cut out the measured section of the pipe. Attach the valve and fittings to the pipe ends and then turn on the water. Adjust the pressure by turning the adjusting nut at the top of the valve clockwise.

Wear safety glasses or goggles anytime you are cutting pipe, producing chips, or working with sharp edges or chemicals. Also, to protect your hands, wear gloves.

INSTALLING A SHOCK ABSORBER

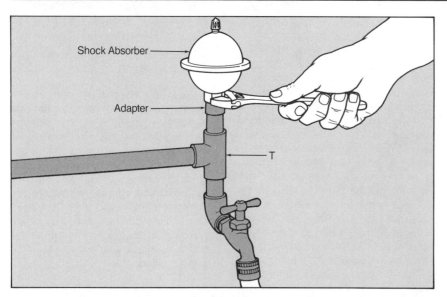

First, read Chapter 3 to gain the skills for this project. To install a shock absorber just ahead of the 'noisy' valve, first insert a T (page 14). Then attach a copper-to-steel adapter to a short piece of copper pipe. Finally, connect the pipe to the T and screw the shock absorber onto the adapter.

Then, periodically check the valve to make sure that it's operating properly.

Pressure-Reducing Valves. A relatively inexpensive and easy way to reduce the water pressure in your home is to install a pressure-reducing valve. These devices effectively drop the pressure from about 80 psi to the more acceptable 50 psi range. Installed in the main line, they are recommended for systems where cavitation is a problem.

Shock Absorbers and Air Chambers. There are two ways to stop water hammer—by installing a shock absorber or an air chamber. Both installations are made at the pipe causing the noise (usually at a clothes washer). Shock absorbers are more expensive but they are smaller and relatively easy to install. An air chamber, basically a capped pipe, can be assembled with standard pipes and fittings; it, of course, will take longer to install.

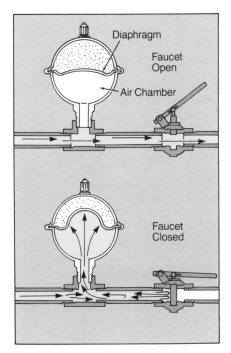

How a Shock Absorber Works. This device will help to eliminate water hammer in your system; it should be installed just ahead of the valve that is producing the unwanted noise. A small device, it consists of an upper air chamber that is enclosed in a flexible diaphragm and a lower 'surge' chamber that accepts excessive pipeline pressure. When the faucet is open, the water flows under pressure. When the faucet is suddenly closed, pressure builds up back along the pipeline. The device absorbs the shock before it becomes water hammer.

CREATING AN AIR CHAMBER

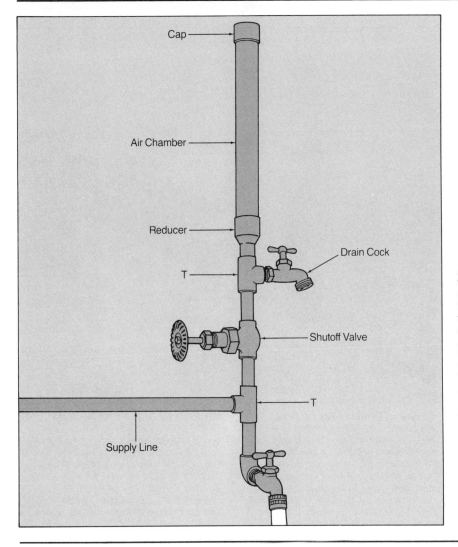

Air chambers, like shock absorbers, are used to control water hammer and they should also be installed just ahead of the valve that is causing the trouble. First, read Chapter 3 to gain the skills for this project. A simple air chamber can be made with standard pipes and fittings; it consists of a capped pipe set into a T near the problem valve. To make a more effective one, as shown here, insert a T to connect the water line to the valve. Add a short piece of pipe and a shutoff valve. To it, add another short piece of pipe, a T, and a drain cock. Above the drain cock, install a short piece of pipe and a reducer fitting. The reducer fitting, on its widest end, should be at least twice the diameter as the water line. To the reducer add a pipe at least 2 feet long and seal it with a cap. Pressure will cause the air chamber to eventually become clogged with water. When this happens, turn off the water at the shutoff valve and open up the drain cock to empty the chamber.

Codes and Permits

Although a National Plumbing Code exists in the United States, its purpose is to oversee the plumbing in federally built structures. There are several national agencies that make recommendations for home plumbing and many states have similar agencies; but ultimately, it is a community's local code or, in some cases, state code that has the force of the law.

Although most home plumbing projects are not inherently dangerous, the results of poor plumbing can be very detrimental to your health and safety. Floods, toxic gas backups, contaminated water supply, electrical shorts, and even explosions can occur when codes are not followed. Local codes dictate the proper methods, materials, and designs that, followed, will help to protect you and your community.

Codes can vary greatly from area to area in regard to materials and methods. What is acceptable and encouraged in one area might be forbidden in another. For example, plastic pipe is prohibited entirely in some communities while in others it is permitted for drain-waste and vent piping only. Likewise, copper pipe, which is acceptable almost everywhere, is prohibited in areas with extremely hard or soft water; in these places, galvanized or plastic pipe must be used.

IMPORTANT

When beginning any plumbing project, check the provisions of your local code. Discuss the project and materials to be used in detail with your local building department. If you are doing repairs you probably won't need to obtain a permit. But for any alterations or additions to your present system, such as the installation of new fixtures or appliances, you should obtain a permit. Failure to do this could result in having to tear out all of your work.

Plumbing and Electricity

If you know anything about electrical wiring, then you know how dangerous it is to work on wiring when water or even dampness is close at hand. Conversely, when working on plumbing, make sure that you will not come in contact with any wiring, electrical pull chains, or similar fixtures.

Plumbing and electricity work together in two additional ways that you should be aware of. One, called *galvanic action,* involves electrical current flowing between pipes; the other, more

dangerous subject is the *ground connection,* the fixture that connects your plumbing system to your wiring system. Familiarize yourself with these two topics and then take precautions to guard against the problems or dangers that they can cause.

Galvanic Action. An electrical current is created when two unlike metals are immersed in water. This *galvanic current* actually transfers the atoms of one metal and deposits them on the other. In home plumbing this process most often causes problems when it occurs between copper and galvanized iron pipe; in time the copper pipe becomes so deteriorated that leaks develop.

IMPORTANT

In order to prevent excessive galvanic action, always install a *dielectric fitting* between copper and galvanized pipes. These fittings have insulators that prevent the two metals from directly touching each other. An option is to use a brass fitting as an intermediate link between the two types of pipe. Neither of these methods will completely halt the galvanic action but they will ensure a longer life for your pipes.

Ground Connections. The cold water piping in your plumbing system, because it is buried, is an excellent conductor of electricity. For this reason, in most homes it is used in the grounding of the electrical system. Called a *ground connection,* a stranded ground wire is run from the main supply panel (circuit breaker panel or fuse box) and is connected by a clamp to a cold water pipe (below). Additional, smaller ground connections might also be visible, for an appliance such as a clothes washer or a telephone system.

WARNING

If you need to work on a pipe that has a ground connection (shown below), you may do so only if you are sure that it is a connection for a telephone or a home appliance. In this case, remove it, do your work and replace it immediately; do not attempt to use the appliance or phone while work is being done. If you know or even suspect that the ground connection is for your entire electrical system, DO NOT ATTEMPT TO REMOVE IT. Either call your local utilities company for advice or contact your local building department.

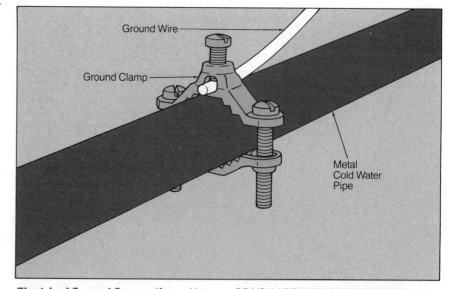

Electrical Ground Connection. Use caution when you see a ground connection like the one shown here. Usually connected to cold-water steel or iron pipes, these fixtures provide the route to ground for an electrical system. If the ground wire is small, it is probably a connection to the telephone system or a home appliance, but if it is large, it is probably the ground connection for the entire house. IF IT IS FOR THE ENTIRE SYSTEM OR YOU ARE UNSURE OF WHAT IT'S FOR—DO NOT REMOVE IT. If it is small, you may remove the ground clamp but while working you should not use any major appliances. In the case of a telephone ground, your phone might not work properly while the clamp is removed. Immediately after doing your plumbing, reconnect the grounding clamp, an essential part of your wiring system.

Preparing to Work

If you are building a new home or adding a new bathroom, then the bulk of your work will be with 'dry' piping—pipes that aren't already connected to the plumbing system. But for most of the projects within this book, the pipes are 'wet', in operation, and must be drained; otherwise you will flood the room that you're working in or, worse yet, you could be scalded by spraying hot water.

Pipes can be drained in two places, at the fixture or at the main shutoff valve. Not all fixtures have their own shutoff valves, and not all separate valves work properly because they are used so infrequently. In these cases you must use the main valve.

Look for fixture shutoff valves under the fixture at the point where the water supply is connected. The shutoff valves for bathtubs and showers are often located behind access panels in the wall or in a closet behind the faucets. Before working on a fixture, shut off the water by turning the shutoff valves. Then turn the fixture faucets on and completely drain the pipes. Shutoff valves are sometimes inoperable; in this case, turn off the water at the main shutoff valve.

Learn where all of your valves are located and check to see if they are in operation. This knowledge can be especially helpful in a plumbing emergency.

SHUTTING OFF WATER

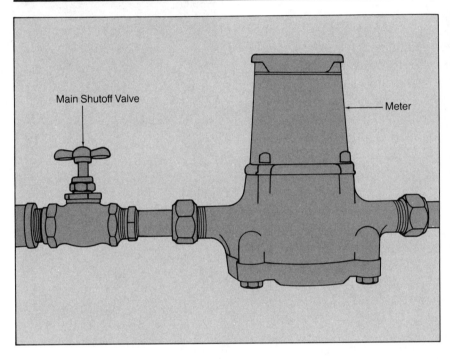

Main Shutoff Valve

Meter

Shutting Off Water at the Main Shutoff Valve. When working on general piping or at a fixture with no shutoff valves of its own, you must turn the water off at the main shutoff valve. These are usually located between the house foundation and the water meter—or, in cold climates, directly inside the foundation wall in the crawlspace or basement. After locating what you think is the main shutoff valve, you should test it to make sure. If the valve itself does not work properly, you should call your water company; they will provide the special servicing needed to shut off the water.

2

Plumbing Tools and Materials

If you've ever been to a plumbing supply store, then you're familiar with the endless racks and bins of pipes, fittings, and tools. Such a spectacle can quickly dampen your hopes of becoming a successful home plumber. In this chapter, however, you will learn only the most basic tools and materials of the trade.

Pipes and fittings are the primary materials used in plumbing. The four different kinds of pipe will be described as well as a brief history of their use in homes. As you'll see, the plumber's language is simple because it is very 'visual'. For example, an 'elbow' is just that—a fitting with a bend in it. Luckily too, these easy names of fittings are the same whether the material is steel or iron, copper or plastic.

As mentioned in the previous chapter, the type of materials you can use are determined by local code. Since many of the tools in the tool set are for use with copper pipe, it would be wise to check your code before going out and purchasing all of them. There has been quite a strong leaning toward the use of plastic pipe in recent years and if your code permits its use you should definitely take advantage of this easy-to-use material.

In truth, the home plumber can do wonders with common household tools. Repairs can be made with a screwdriver, a moderately-sized adjustable wrench, and an ordinary pair of pliers. Depending on what your intentions are, you might not need to know about all four kinds of pipe. Still, reading the entire chapter will lend some mental order to your otherwise puzzled image of the plumbing supply store...and it will build your confidence for the next chapter on skills.

Basic Plumbing Tools

Plumbing tools are neither complicated nor expensive. However, they are specialized, and having the correct tool handy for a particular job is certainly a wise practice, especially when the job is the result of an emergency. The tools shown on the next page are very basic. They are for working on the most common types of fixtures and on the type of pipes that are prevalent in most homes. They will aid you in emergencies such as eliminating clogs, and also in general repairs and installations. Before acquiring all of them, however,

you should check the plumbing in your home for the kind of pipes used; some tools shown here might not be relevant to your system.

Purchase your tools at a plumbing supply store or at a well-stocked hardware store. When different brands are available, select the best brand, especially when it's a tool that you'll use often. Good tools always pay for themselves—by cutting down on 'slipping accidents' and related aggravations. Especially with tightening and loosening tools, the design and quality is important. Even with big arm muscles behind it, a poor tool will make the project a pain.

All of the tools in this basic plumbing kit should cost a little over $100.00 which isn't really a huge investment when you consider the cost of hiring a plumber to do the work for you. If you have tools in your tool kit right now that you think might be useful in plumbing projects, you might want to inspect them carefully to make sure that they're similar to what we show here.

In addition to the tools shown, there are tools used for some of the projects in this book that are so standard that we've elected not to present them here. These include: screwdrivers (slot and Phillips), an adjustable-end wrench,

TOOLS FOR HOME PLUMBING

1 Plunger. Also called a *plumber's helper,* this tool is a must for unclogging drains. Although there are other simpler, less expensive types, the one shown here with the conical base is the only kind that is effective for toilets. The bottom part can be folded up when it is used on flat drains.

2 Snake. Actually a *trap and drain auger,* this tool is used for clearing out traps and branch drains. Usually made of ¼-inch twisted steel wire, they measure from 10 to 12 feet long. On the end is a corkscrew-like head that tears through dense matter. One model, called a *spin auger,* is inexpensive and has an easy-to-use crank. For unclogging main drains, use a longer and larger ½-inch snake. If that doesn't work, consider renting an even larger snake with an electric motor or call for professional help.

3 Pipe Wrenches. Sometimes called *Stillson wrenches,* these tools are the workhorses of the plumbing trade. You'll need two—one measuring 12 to 14 inches from top to bottom for use on supply pipes, the other 18 inches for use on waste pipes. Often the two are used together with one wrench holding and the other doing the turning. When wrenches are used to grip exposed brass or chrome-plated piping, cover the teeth with tape or rags.

4 Spud Wrench. Made with wide toothless jaws; used for tightening and loosening the large nuts on toilets and sinks.

5 Basin Wrench. A specialized wrench that easily gets to nuts located behind sinks and in other tight places.

6 Valve Seat Dresser. For use on compression type faucets; used to grind and smooth non-replaceable seats.

7 Valve Seat Wrench. Also used on compression type faucets, to remove valve seats that are beyond repair.

8 Tube Cutter. For use with copper tubing; has a built-in reamer that scrapes burrs off the cut edge.

9 Tube Bender. For bending copper tubing more accurately and with less effort than by hand.

10 Propane Torch. Used for soldering copper tubing; shown with a flame-spreader attachment which is helpful when thawing frozen pipe.

11 Solder. Solder comes in a solid wire form; use it with a torch when sweating (joining copper tubing). Available with varying percentages of tin and lead.

PLUMBER'S TIP: There have been recent problems with higher than recommended levels of lead showing up in drinking water in new buildings. Some local jurisdictions have been banning high-content (50 percent tin/50 percent lead) solder. Check with your local building department.

12 Emery Cloth. For smoothing the edges of copper tubing when sweating; a good substitute is steel wool.

13 Flux and Brush. Flux is applied to copper tubing to prevent it from oxidizing when heat is applied; the special brush makes application easier.

14 Sheet Metal. Use metal sheeting if you are soldering within 6 inches of flammable material such as studs or joists.

15 Flaring Tool. Use this tool to connect copper tubing without sweating; it flares the ends of pipes so they will accept special fittings. Also can be used to join CPVA and PB plastic pipe to metal pipe.

16 Fire Extinguisher. Keep this firefighting tool handy in case of emergency.

a ball-peen hammer, rib-joint pliers, a measuring tape, a wire brush, a miter box, and a flashlight or trouble light. As part of your gear, also have a pair of safety glasses or goggles, a necessity for many plumbing projects, and a pair of work gloves.

Depending on your local code, you might be able to use plastic pipe for many installations. If so, you'll need tools specifically for that purpose (page 15). If you want to tackle projects that involve the use of cast iron pipe, you should attain (by purchasing or renting) tools for these special jobs (page 17).

Once you have bought your new plumbing tools, try to keep them in a plumbing tool box or in some way separate from your general tools. Having them organized will make for a much smoother operation in the event of an emergency. You might also want to assemble an 'emergency plumbing kit' specifically for emergencies, (page 36). With it and your basic tool kit, you'll be well-supplied when things go awry with the pipes.

Choosing the Right Materials

Pipes and fittings are the elements of any plumbing system. Other materials used in a system are the substances to hold the pipes together and the devices to hold the pipes in place. In some cases you will have no choice of which materials to use; the code will dictate that only a certain kind of pipe be installed for a particular purpose. In other cases, you will have a choice. One important point to remember is that if you combine different types of metal pipes, you must use special fittings between them (page 33).

When choosing materials, cost is always a consideration, both of the piping and of the tools needed to work with it. You should also consider, however, the skill level required and the subsequent time involved in working with each kind of material. Familiarize yourself with plumbing skills (pages 18-33) and then make your choice. Generally, it's best to choose the material that's easiest to work with.

Galvanized Pipe and Fittings

If your home was built in the 1950's or earlier the chances are great that you have galvanized supply pipe. The main component of such pipe can be steel or iron, or, in rare cases, brass. The term 'galvanized' refers to the coating of zinc on it, a process intended to make the pipe more rust-resistant. Ironically, galvanized pipe is noted for corroding quicker than either cast iron or copper

pipe. Due to its rough interior surface it also tends to collect mineral deposits thereby impeding the flow of water and, in some cases, affecting the water pressure.

A major characteristic of galvanized pipe is the threading by which pipes are connected to each other. (Sometimes it is called 'threaded' pipe for this reason.) Threading can be on the exterior or interior of pipes. In the plumbing trade, the 'gender method' of describing pipes is used (i.e., male/female). Connecting threaded pipe is easy but once it's installed it cannot be taken apart without disrupting the system. The problem is that when you unscrew one end of the pipe you are simultaneously tightening the other end.

If you need to repair galvanized pipe, instead of removing a threaded piece, you'll have to cut out the faulty section. In this case, it's best to replace it with galvanized pipe and a *union*, a fitting that saves you the trouble of turning pipes. But if you're extending a supply system composed of galvanized pipe, use copper pipe for the run, or, if your code permits it, plastic.

Galvanized pipe comes in many sizes and lengths. The standard sizes used by homeowners are ½-, ¾-, and 1-inch diameter although larger sizes are also available. Standard lengths are 10 and 20 feet. Smaller lengths, called *nipples,* are easily obtainable in precut and pre-threaded sizes. If you need special sizes of threaded pipes, it's best to let your supplier thread them for you. Doing it yourself involves buying or renting special equipment—a vise and a pipe threader.

Fittings are used where pipe changes size or direction or, in the case of the cap, stops altogether. Shown on page 14 are only the basic types of fittings used in the home; a much wider range is available. Additional materials used with galvanized pipe include teflon tape and pipe joint compound, both used to seal pipe together.

Copper Tubing and Fittings

Copper pipe, called 'tubing', is more costly than galvanized pipe but its benefits are considerable. It's tough, highly resistant to corrosion, and has a smooth interior surface that allows for an easy flow of liquids. It's much easier to work with because it's lighter and because assembly goes faster. Flexible copper tubing can be simply bent, saving you the cost and work of putting in a lot of fittings.

GALVANIZED STEEL PIPE AND FITTINGS

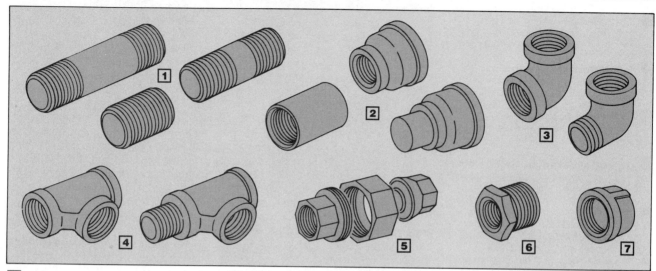

1 Nipple.   A precut pipe threaded on both ends. Usually available from 1½ inches to 6 inches in ½-inch increments or from 6 inches to 12 inches in 1-inch increments. A 1½ inch nipple threaded on its entire length is called a *close nipple*. A similar length with a very short unthreaded section is referred to as a *shoulder nipple*.

2 Coupling.   Couplings join two pieces of pipe in a straight run. They usually have female openings at both ends. *Straight couplings* have the same size openings at each. A *reducing coupling* has a smaller opening at one end. A coupling with a female and a male opening is called a *reducing piece*.

3 Elbow.   Also called an *ell*, this fitting is used when the pipe changes direction. Available in 45° and 90° angles with both openings usually female. An elbow with female and male ends is called a *street ell*. Reducing ells are also available.

4 T.   Used when a pipe continues in the same direction and also branches in a right angle. A *street T* has two female openings and one male end. Reducing Ts are also available.

5 Union.   A union joins two lengths of pipe of the same diameter. Unlike a coupling, it can be installed and removed without turning either pipe. It consists of three parts: a

male thread piece, a shoulder piece, and a nut in the center that connects the two together. Unions are most often used where appliances or fixtures are periodically replaced, such as at a water heater.

6 Bushing.   A fitting that reduces one opening in a tee, elbow, or other fitting so that it can accommodate a smaller pipe. It is threaded both inside and out and is hexagonal on one end so that it can be turned with a wrench.

7 Cap.   A solid piece of galvanized steel, threaded inside. Used to seal the end of a pipe as when making an air chamber or closing off a pipe no longer in use.

COPPER TUBING AND FITTINGS

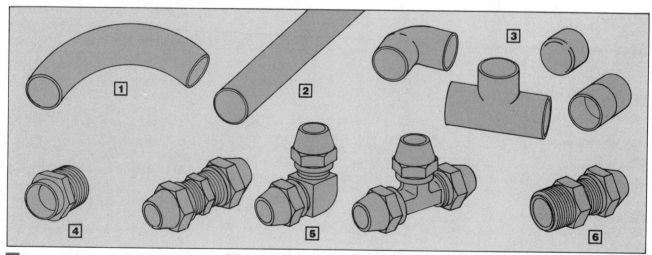

1 Flexible Copper Tubing.   This soft pipe can be easily bent and shaped as needed. Available in straight lengths of 20 feet or in coils of up to 100 feet.

2 Rigid Copper Tubing.   Hard pipe that cannot be bent; with it you need to use elbows to turn corners. Available in 10- and 20-foot lengths.

3 Solder Fittings.   These fittings (the ell, coupling, tee and cap), although they are soldered together, function in the same way as do their counterpart galvanized steel fittings (above).

4 Adapter for Soldering.   Copper fittings are available for soldering on one end and fitting with threaded pipe on the other.

5 Flare Fittings.   These fittings (coupling, ell, and T) are all used when soldering isn't possible and tubing is flared together (page 26).

6 Adapter for Flaring.   A fitting for flaring on one end and joining to threaded pipe on the other.

There are several types of copper pipe and they are classified according to how they are used. Supply pipe comes in two forms: flexible (soft) and rigid (hard). A third kind of supply pipe (not shown here) is the small-diameter corrugated tubing that is often used to link up hard or soft supply pipes to fixtures. Although copper pipe is most commonly used in supply lines, there is yet another kind, in larger diameters, that is used for drain-waste and vent systems.

Flexible Copper Tubing. Flexible tubing is more expensive than the rigid variety but it offers some worthy advantages. It can be bent by the homeowner to go in whatever direction is necessary. The need for fittings is reduced—a factor to be considered if you're making many turns or working in tight places. Additionally, flexible copper tubing doesn't require sweating (soldering). Instead, pipes may be connected with flare joints or compression joints. This can not only be a time-saver but it's also a much safer method if you're working in an area that could easily catch fire from soldering.

Sold in 30-, 60-, or 100-foot coils or in straight lengths, flexible tubing comes in two weights: medium, or Type L and heavy, or Type K. The latter is used mainly for outdoor, underground lines. Various diameters of pipe are available. Unlike other kinds of pipe, they are all sized by the inside diameter, not the outside.

Rigid Copper Tubing. Rigid tubing is obviously tougher but it isn't quite as easy to work with as the flexible type. Still, it is preferable to galvanized pipe. Sold in 10- or 20-foot lengths, it is available in three weights. Two heavier-weight types, Type K and Type L, are primarily used underground. The lighter-weight Type M, used above the ground, is generally acceptable for home plumbing. Various diameters are available.

Copper Fittings. Copper fittings are similar in name and shape to galvanized fittings. Some of them are sweated on; others are joined by flaring. Solder is the substance used for joining pipes. It and the necessary tools for cutting, bending, and soldering, are shown on page 12.

Plastic Pipe and Fittings

The most modern of all piping materials is plastic and the rewards of using it are numerous. In fact, it far surpasses the other materials on almost all counts. It is less expensive to use than metal, is lighter, is easier to install, and requires very few tools. Like copper, it has a

PLASTIC PIPE AND FITTINGS

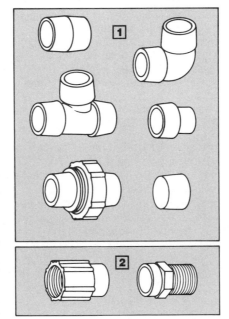

smooth interior surface that allows for the easy flow of liquids. But unlike metal piping, plastic pipe is resistant to rust, damage from chemicals, and corrosion due to galvanic action. Last but not least, if you live in a cold-weather area, plastic pipe will give you far fewer worries about freezing and bursting pipes.

Local codes are becoming more and more receptive to the use of plastic pipe but some codes still restrict its use. The reasons for this are varied. Because it is a relatively new material, some people feel that it hasn't been tested for a long enough period. Certain types of plastic pipe become limp when hot water (over

■ **Plastic Pipe.** Plastic pipe (not shown) is available in several types. It should be chosen for the use intended. **PVC.** Stands for *polyvinyl chloride*. A rigid white or light-colored plastic pipe; often used for DWV systems, sometimes used for cold-water supply lines. **CPVC.** *Chlorinated polyvinyl chloride pipe*. A rigid white or light-colored plastic pipe; used for hot or cold water runs. **ABS.** *Acrylonitrile-butadine-styrene*. Black or gray rigid pipe used almost exclusively for DWV systems. **PB.** A newer form of flexible plastic pipe, *polybutyline*, also colored black or gray. Originally used by water and gas companies in irrigation systems, it is gradually being accepted for use in supply lines of home plumbing.

1 Plastic Fittings. These fittings (coupling, ell, T, bushing, union, and cap) all function like their galvanized steel pipe counterparts (left, top).

2 Plastic Adapters. Adapters are available for joining one end to plastic pipe and the other to threaded pipe.

140°) is run through them. Similarly, most plastic pipe is more flammable than metal in case of a fire. Yet another reason is that labor unions fear a loss of employment if plastic pipe becomes the norm, since it requires fewer skills to work with.

If your local code permits the use of plastic pipe, by all means take advantage of the proposition; but if its use is forbidden, don't even consider it. Always work to code; otherwise you may cheat yourself and have to tear out all of your work.

Plastic pipe is joined by a permanent solvent-cement or by compression fittings. Joints are just as secure as

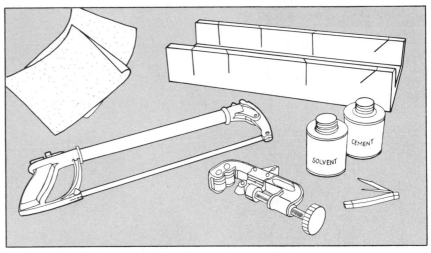

Tools For Working With Plastic Pipe.
These tools are necessary for working with plastic pipe. Use either a special tubing cutter (one made only for plastic pipe) or a fine-toothed hacksaw for cutting. If you use a saw, make sure that your cuts are straight by using a miter box. Keep a pocket knife and/or sandpaper handy for smoothing off edges. Solvent-cement comes in different varieties according to what kind of pipe you are cementing. The brushes are included as part of the container.

those made for metal pipes and the type of adhesive is determined by the type of plastic. Since plastic pipe is a relatively recent innovation, you probably won't find any in your home that needs repairing. If you do, removal must be made by cutting. Plastic pipe requires special tools and materials as shown on page 15.

There are several types of plastic pipe and they are divided into two categories: flexible and rigid. The most common flexible pipe is PB, sometimes used for hot and cold supply lines. Of the rigid types, there are three varieties that are widely used. PVC is used in DWV (drain, waste, and vent) systems and cold water supply lines. CPVC is used for cold or hot supply lines and ABS is used in DWV runs.

Your local code will recommend what type of pipe you should use for a particular project. Most codes restrict the connecting of one type of plastic pipe with another. The code will also specify what *schedule* the pipe must have. The schedule of the pipe is the pressure rating; it will be printed directly on the pipe. Because plastic pipe is so lightweight, it doesn't withstand line surges as well as metal pipe. Besides using pipe with the correct pressure rating there are other precautions that you should take to alleviate pressure problems. Install air chambers at fixtures if necessary (page 7) and, more importantly, adjust your water heater.

Flexible Plastic Supply Pipe. Unlike other kinds of plastic pipe, the flexible kind cannot be joined with an

WARNING

If you use PB or CPVC plastic pipe for hot water supply lines, be sure to adjust the temperature and pressure-relief valve of your water heater (page 112). Set it to match the temperature and pressure rating of the new pipe. The water temperature should be no higher than 180° which is generally too hot to be safely usable anyway.

adhesive. Instead it must be joined with compression fittings like the ones that are used to join copper tubing (page 27). But because it may be bent, flexible pipe can follow a winding route without the use of a lot of fittings. For this reason, it is especially suitable in cramped places. Flexible pipe also may be joined to other materials with the use of transition fittings (page 33). It comes in rolls of 25 and 100 feet and can often be cut to the length you need.

Rigid Plastic Supply Pipe. Even though they are termed 'rigid', both PVC and CPVC pipe are slightly limber and can accept minor changes in direction without cracking. Rigid pipe is joined with solvent-cement which must be purchased to match the kind of pipe you are using. Fittings for this pipe are similar in design to those used for metal pipes. Transition fittings allow you to connect it to other kinds of pipe. If used for hot water lines, the method for hanging rigid plastic pipe is different than that for metal pipe; it should be clamped at 3-foot intervals and hung loosely to facilitate movement. Additionally, plastic

supply lines should be offset every 10 feet of a run with a 12-inch offset for expansion and contraction.

When shopping for supply pipe for indoor use, look for the National Sanitation Foundation stamp of approval. Rigid plastic pipe comes in 10-foot and 20-foot lengths and can be bought in single lengths or in bundles. Diameters are the same as those of metal pipes, the most common being ½-inch, ¾-inch, and 1-inch sizes.

IMPORTANT

When storing rigid plastic supply pipe, do not leave it in direct sunlight for more than one week. Ultra-violet rays from the sun will cause it to become brittle.

Rigid Plastic DWV Pipe. PVC and ABS are both easier to connect and to hang than cast iron pipe. However, PVC is the preferred of the two because it is less susceptible to chemical damage and is resistant to fire. Its use also guarantees you a greater variety of fittings. DWV fittings (also called *sanitary fittings*) vary according to what materials are being joined. Like cast iron fittings, they have no shoulder to block the flow of wastewater.

Plastic DWV pipe is usually sold in lengths of 5, 10, and 20 feet. Size (diameter) varies according to use. Vent pipe sizes range from 1¼ to 4 inches; toilet and house drainpipes are available in 3- and 4-inch diameters. Drainpipe for lavatories, sinks, and tubs measure either 1½ or 2 inches in diameter.

CAST IRON FITTINGS

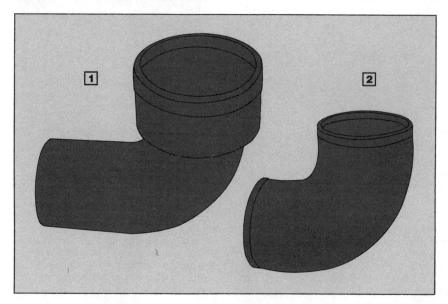

Cast iron fittings come in a great variety of shapes and angles similar to other types of piping (tees, offsets, and Y branches) but the novice home plumber usually will not tackle a complicated run of cast iron pipe. Instead of being called 'ells' like other types of piping, angled pipes are called *bends*. In making plans for new plumbing that involve cast iron pipe, keep the runs simple—with as few fittings as possible.

1 Hub Fittings. Also called *bell and spigot* fittings, these pipes are joined with oakum and molten lead; often seen in older homes.

2 Hubless Fittings. These fittings are much easier to work with since they are joined with special clamps and rubber sleeves. They also take up less space and may, in some cases, fit within a 2 x 4 partition space, unlike most hub fittings. Check with your local building code to make sure that these are an acceptable material.

Cast Iron Pipe and Fittings

Most American homes have drain-waste and vent pipes made of cast iron. Cast iron pipe is heavy and strong. It's resistant to corrosion and contains noise well. But because of its weight, it is difficult to work with. Homeowners rarely work on the cast iron pipe in their plumbing systems and if they do, it is usually to add a fitting. If your code allows it, make any replacements in a cast iron system with plastic pipe.

If your code won't permit the use of plastic pipe then the next best thing to use is hubless fittings. At one time, cast iron pipes were joined with oakum and molten lead (hub fittings). Today, the easier-to-use hubless fittings are the norm; most codes permit them but, as always, check for restrictions before making your purchase. Cast iron pipe requires special tools not found in the basic tool set on page 12.

There are two grades of cast iron pipe available—service weight and extra-heavy weight. Pipe and fittings will be stamped—either *S & V* for service weight or *XH* for extra-heavy. If the code allows it, use service weight.

Like plastic DWV pipe, cast iron pipe can be purchased in lengths of 5, 10, and 20 feet. Available diameters are 1½, 2, 3, and 4 inches.

Pipe Support Devices

Besides the pipe itself, another essential element of any correctly installed plumbing system is the support system—the devices that hold the pipe in place. Devices are chosen according to what kind of pipe is being supported and where it is being hung or supported. Local codes should supply this information.

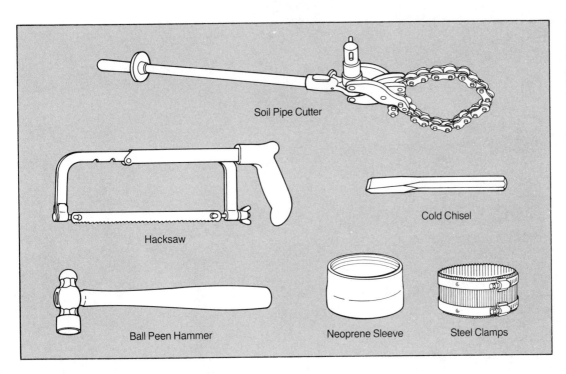

Soil Pipe Cutter

Hacksaw

Cold Chisel

Ball Peen Hammer

Neoprene Sleeve

Steel Clamps

Tools For Working With Hubless Cast Iron Pipe. These are the tools needed for working with hubless cast iron pipe. Cut already-installed pipes with a soil pipe cutter, normally available at rental stores. For installing new pipe, use a hacksaw, a ball peen hammer and a cold chisel. Materials needed include neoprene sleeves and steel clamps.

DEVICES FOR SUPPORTING PIPE

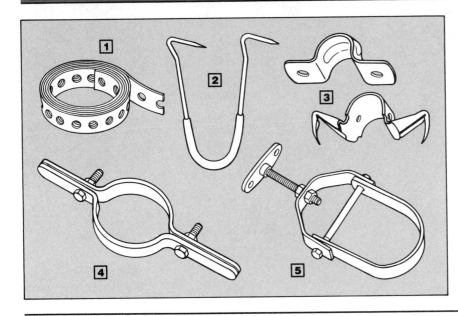

1 **Plumber's Tape.** Made of copper, galvanized steel or plastic; to be used with the corresponding kind of pipe. Nails are required.

2 **Wire Hanger.** Used for metal pipe, self-nailing. Shown is an insulated (plastic-coated) hanger.

3 **Pipe Straps.** Used for galvanized pipes. Two types are shown—one requires nails, the other is self-nailing.

4 **Floor Clamp.** Supports vertical pipes, such as a stack, within a wall.

5 **Metal Hanger.** Supports steel or cast iron pipe. Similar hangers are made of plastic for plastic piping.

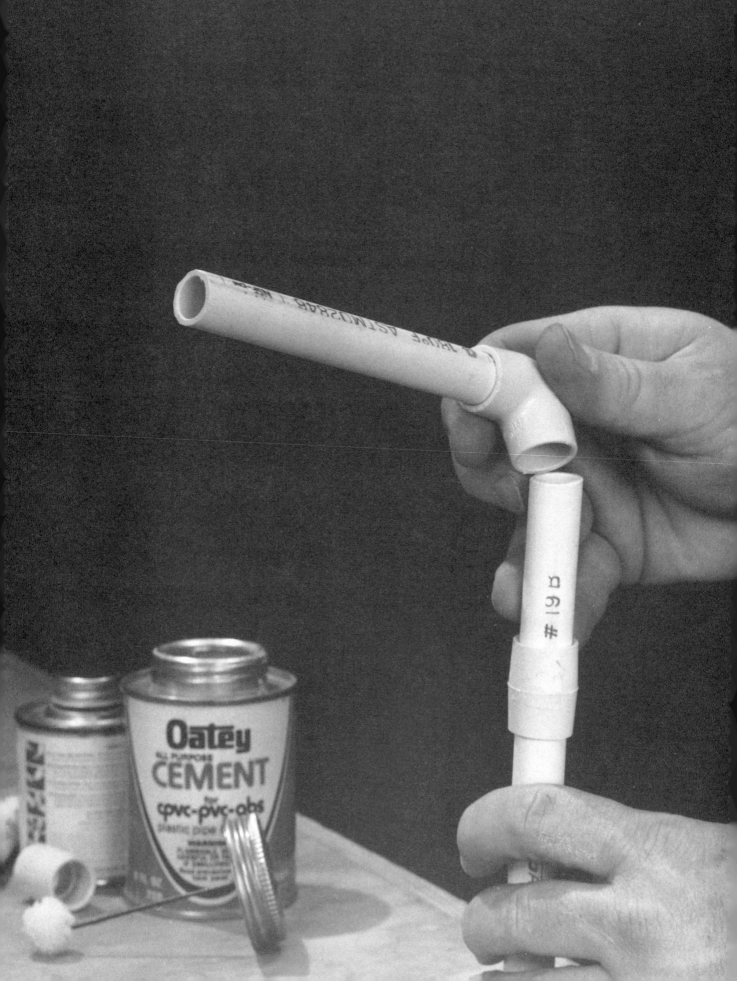

Basic Plumbing Skills

This chapter on skills relates hand-in-hand to the previous chapter on materials. There you learned about the four basic kinds of pipe. Here you will learn how to work with those pipes and how to put plumbing tools in motion.

The primary activity of a plumber is pipefitting. This involves measuring, cutting, shaping, and assembling pipes and fittings. Though all are equally important, joining is the one people think of most often when they picture a plumber at work. Technically, there are three ways of joining. The first is to screw pipes together by matching threads on their ends. Second, there are fused joints; these are made by soldering or cementing. Last, there are compression joints. Basically these are unthreaded pipes held together by threaded nuts—such as the flare joints used in copper tubing.

Galvanized, copper, plastic, and cast iron pipe will be presented with directions for how to handle, cut and join it. Attention is given to pipe repairs and the standard rules for supporting pipes. Although each kind of piping calls for a different process, there are also some general practices that apply to all kinds of pipes. Measurements of pipe diameters are always taken on the inside of the pipe—not the outside. Also, a professional plumbing tip that can save you neck cramps and eye strain, is to assemble pipe, as much as possible, at a workbench.

None of these skills is difficult—sweating, using solvent-cement, even working with hubless cast iron pipe can all be done correctly if done methodically the first time. They can all be mastered with minimum practice. Most importantly, familiarize yourself with the safety precautions for each skill.

Working with Galvanized Pipe

Galvanized pipe (page 14) is the trickiest of all kinds to work with. Because it is threaded to fit precisely into a plumbing run, you must be extremely accurate when taking measurements.

A Word About Threading. The threading on pipes used in plumbing is uniquely tapered so that on the end of a pipe it is smaller in diameter than it is ½ inch further up the pipe. This special feature causes a joint to seal as the pipes are screwed together and tightened. The threads of galvanized pipes are standardized so that you never have to worry about them not matching. In fact, the only other kind of threading that you will find in a plumbing system are the threads on compression fittings, the brass tubing used for sink drains, and ends of faucets for garden hose hookups.

When galvanized pipe was more popular, home plumbers used to thread their own pipes. This process requires a considerable investment in equipment—a special vise for holding pipe and a pipe threader. Not shown here, these two pieces of equipment may be rented when making large installations of galvanized pipe. For smaller installations, you should have pipe cut and threaded at a plumbing supply store; but be sure to measure accurately to keep your travelling time to a minimum. When you receive your pipe, it will have straight couplings on the threaded ends. These are for protection of your hands as well as the pipe threads. They may also be used, if necessary, as long as pipe-joint compound is used to seal the joint.

Measuring, Cutting, and Disassembling Galvanized Pipe. The least expensive way to replace a length of leaky galvanized pipe is with the same material. Galvanized pipe is, furthermore, accepted by almost every code while copper and plastic are not as universal. Unless your leak is at a pipe run that is joined by a union, you will have to disassemble the pipes by first cutting them. Then, you can remove the damaged pipes with two wrenches as shown in the illustration.

Once galvanized pipe is joined it cannot be easily removed. The problem is that as you unscrew a pipe at one end, you will be tightening it at another. For the same reason, when you make a replacement, you must use a union. A union consists of three parts: a piece at each end that turns onto the pipe and a large 'nut' at the center that pulls the two end pieces together. No pipe-joint compound is needed where the two end pieces meet each other, but compound or fluorocarbon tape is recommended where the two end pieces meet the pipe and where the nut is tightened.

Measuring is critical and you should allow for all fittings. Also, it is better to get just a tiny bit more pipe than what is needed since you can usually tighten

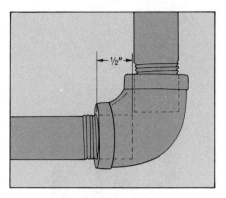

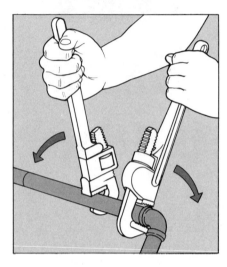

Measuring Galvanized Pipe. In order to determine the exact size of pipe needed, measure the distance between the faces of fittings. To this figure, add the distances that the pipe extends into all fittings. If you are using ½- or ¾-inch pipe, as shown here, the distance will be ½ inch. If you are using 1- or 1¼-inch pipe, the distance will be ⅝ inch. Note that three threads are visible after the pipe is installed. If measurements are required where no fitting already exists, have someone hold the fitting in place where it will be installed and then measure.

Taking Apart Galvanized Pipe. First, drain the pipe and cut it as described in step 1 below. Wear gloves and eye protection and use two wrenches for this process. With one wrench grip the pipe next to the one you're removing; hold it with a steady pressure while turning the other wrench in the opposite (counterclockwise) direction. The open jaws of the wrench should be pointed in the same direction that you are turning. If the pipes are resistant, apply a liberal dose of penetrating oil to the joint and wait about five minutes before trying again.

REPLACING DAMAGED GALVANIZED PIPE

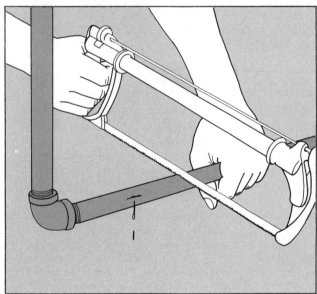

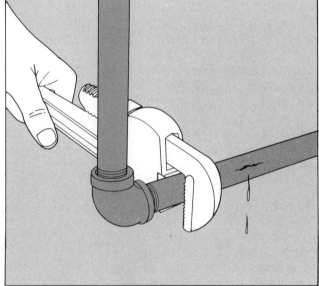

1 If the pipe is leaky, first try other methods of stopping the leak (page 37). If these fail to work, you must replace the pipe. In this instance the pipe is at least 6 inches long—long enough to use a nipple and a union. If you were to unscrew the pipe at one end you would be tightening it at the other so the only way to remove the pipe is to cut it. (The only exception to this is when

there is a union within the run; then you could begin loosening the pipe there.) With the water turned off at the main and run out of the system, place a bucket under the leak to catch any remaining water. Then steady the pipe run with a firm grip or a wrench as you begin sawing. Use a coarse-toothed hacksaw for the job (18 teeth per inch) or, better yet, a pipe cutter. Cut the pipe at a right angle.

2 Next, unscrew both pieces of cut pipe. One pipe will be discarded and replaced with a nipple. The other pipe can either be replaced with a pipe of the same size or it can be saved, recut, and rethreaded at a plumbing supply store. In any case, measure for it, the nipple, the union, and all threading.

ASSEMBLING GALVANIZED PIPE

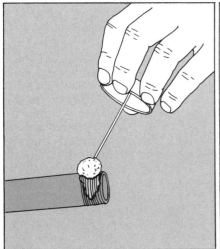

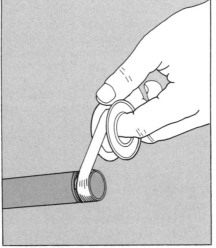

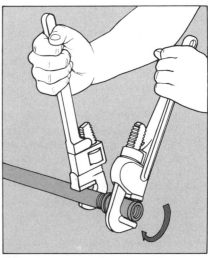

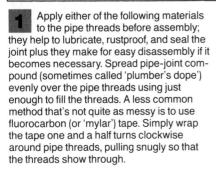

1 Apply either of the following materials to the pipe threads before assembly; they help to lubricate, rustproof, and seal the joint plus they make for easy disassembly if it becomes necessary. Spread pipe-joint compound (sometimes called 'plumber's dope') evenly over the pipe threads using just enough to fill the threads. A less common method that's not quite as messy is to use fluorocarbon (or 'mylar') tape. Simply wrap the tape one and a half turns clockwise around pipe threads, pulling snugly so that the threads show through.

IMPORTANT

Always use sealing materials on pipes or fittings with exterior threading; do not use on interior threading.

CAUTION

Any further tightening could strip the threads and result in a leaky pipe.

2 Begin by hand-assembling the fitting to the pipe to make sure that they are not cross-threaded. Then, with the pipe held stationary with one wrench, turn the other wrench clockwise to tighten the joint. The open jaws of the wrench should be pointed in the same direction that you are turning. The pipe is completely tightened when exactly three threads remain visible.

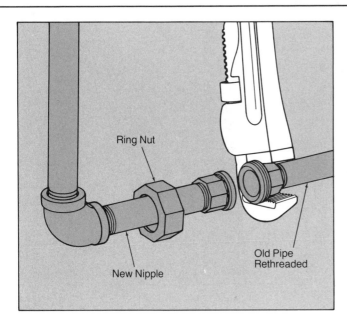

Ring Nut

New Nipple

Old Pipe Rethreaded

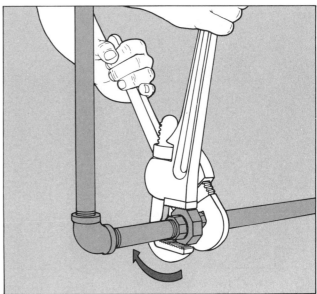

3 Begin the assembly by preparing the old (or new) pipe threading with compound or tape and screwing one of the union fittings to it. Slip the ring nut over the nipple and then prepare its threads for joining at both the existing fixture and the new union.

4 Connect the union. The faces of the union nuts should touch. If they do not you must begin again using a slightly longer nipple. Slide the ring nut to the center of the union and screw it by hand. Then tighten it with a wrench while holding the exposed union nut with another wrench.

the pipe by one more turn. If the pipe is still too long, you can usually have it rethreaded; but, if it is too short, it must be discarded.

Assembling Galvanized Pipe. When adding new or replacement pipe in a system, it is best to use either copper tubing or plastic pipe. But if your code permits neither kind or for some other reason you wish to use galvanized pipe, you can do so. Remember, however, that the pipe will be heavy and that you will have to measure very carefully.

A method that will help you in assembly of this kind of pipe, and all kinds for that matter, is to preassemble as much as possible at a workbench. This technique, used by professionals, will save you a lot of needless straining with wrenches; it is foolproof as long as you continually check your measurements.

Use two pipe wrenches to assemble galvanized pipe; one to hold the 'last-assembled' pipe in place, the other for turning the 'next' pipe. Pipe wrenches are designed to grip pipes tighter as the wrench is turned. Avoid putting tremendous pressure on a pipe as you might cause the wrench to slip off and at the same time scratch the galvanized zinc surface, the rust-preventive finish, off the pipe.

Galvanized Pipe Repair Skills. Two methods of repairing galvanized pipe are shown. One shows what to do when the leak is in a length of pipe long enough to be replaced by a nipple and a union. The other method shows what to do when the length of pipe is too short to accept a nipple and a union. In the latter case, you must simply 'move ahead' to the next pipe that is long enough and install the nipple and union there instead. Galvanized pipe should be supported by hangers every 6 to 8 feet in horizontal runs and 8 to 10 feet in vertical runs.

A common problem in repairing galvanized pipe is the prevalence of corrosion. When you remove a pipe and discover lime or a similar deposit thickly lining the interior, the best thing to do is to replace as much of the pipe as possible. Besides causing leaks, such corrosion also reduces the volume of water flowing through the pipe within your home. Remember, when connecting galvanized pipe to copper tubing, you should use a dielectric union between the two to prevent galvanic action (page 33).

Working with Copper Tubing

Copper pipe (page 14) is still the most popular type for supply lines; it can also be used for drain and vent lines although its relatively high cost is prohibitive. Flexible tubing is preferred since it is easier to work with and can be joined by compression or flare joints. If your local code specifies rigid tubing, you'll have to assemble it by soldering.

Measuring, Cutting, and Bending Copper Tubing. To measure for copper tubing, first add the distances of piping between all fixtures. To this figure, add the 'pipe-insert' distances—the distance that pipe is inserted into all fittings. Since this distance will vary according to the fitting used, use the drawing here as a guide and then, with all your fittings handy, preassemble pipe to fittings and measure.

Copper tubing should be cut with a pipe cutter fitted with a copper-cutting blade. If you're a novice, before doing your actual project, you might want to practice cutting some scraps, just to get the feel of the tool. An optional cutting method is to use a fine-tooth hacksaw and a miter box. Once cut, burrs on the pipe should be removed with the retractable reamer on the pipe cutter or with a file. Also file or sand off any outside burrs.

Rigid copper tubing requires the use of fittings to change the direction of a pipe run. Flexible copper tubing, however, can be simply bent with bare hands or with the aid of a tube bender.

REPLACING A SHORT RUN OF GALVANIZED PIPE

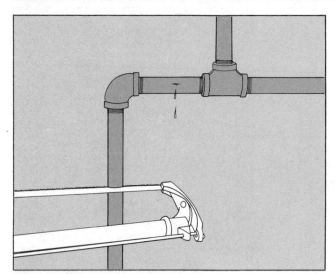

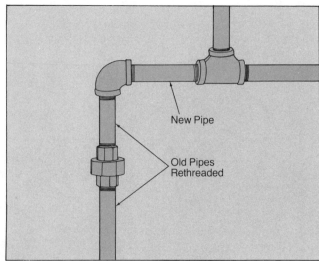

New Pipe

Old Pipes Rethreaded

1 In this example, the damaged pipe is not long enough to be replaced by a union and a nipple. The solution is to install the union in the nearest pipe that is long enough. Turn off the water and drain the pipe. Cut the nearby pipe with a hacksaw. Disassemble and replace the damaged section with a new pipe of the same size. Salvage the pipe that you have cut, get a small section cut out of one end and have both ends threaded to accept the union.

2 Assemble all pieces as shown, beginning with the new pipe and using a sealing material for joints. Slip the union nut over one pipe and, finally, connect the union as in step 4, page 21.

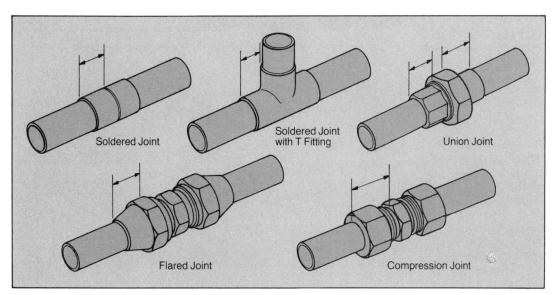

Measuring Copper Tubing. Shown are pipe-insert distances for common copper fittings. When measuring, add these distances to the total length of pipe between fittings.

Soldered Joint

Soldered Joint with T Fitting

Union Joint

Flared Joint

Compression Joint

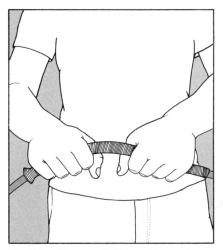

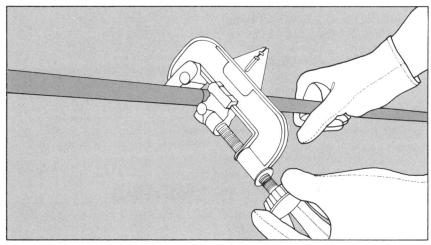

Bending Copper Tubing. The easiest way to bend copper tubing is to use a tube bender. Simply slip tubing into it and apply pressure until it is bent the way you want it.

Cutting Copper Tubing. To cut copper tubing, first put on gloves and eye protection. Next, open the cutter far enough to slip in the tubing. Then place the blade directly on the measurement mark. With the handle twisted just tight enough to make a shallow cut into the metal, turn the cutter in a complete circle around the tubing. Then slightly twist the handle tighter. Rotate again around the tubing. Repeat again and again until the tubing breaks. A few practices will be sufficient to give you the feel for doing this correctly; twisting the handle too tightly will destroy the tubing.

An optional method is to use a hacksaw; if you use this method be sure to use a fine-toothed blade (32 teeth per inch) and a small miter box.

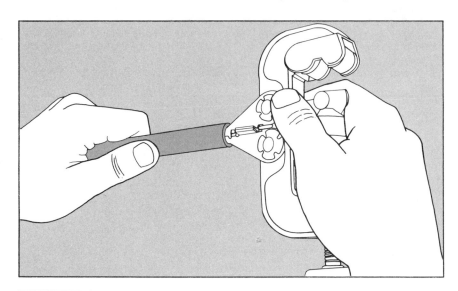

Preparing Copper Tubing for Joining. After you have cut the tubing you must smooth and clean the ends. Often burrs will remain on the outside or inside of the tubing. Most tubing cutters come with a retractable blade for this purpose and you may use it, as shown here, to smooth edges. Files are also useful tools for this purpose, and if you have cut the tubing with a hacksaw you can use a round file to remove burrs from the inside of tubing.

Joining Copper Tubing by Soldering. Because copper is so soft, it will not accept pipe threading; therefore, it must be joined by other methods. Rigid copper tubing must be joined by *soldering (sweating)* or flaring while flexible tubing can be joined by either of these methods or with compression joints. Soldering requires alertness and patience. Before beginning, read and thoroughly understand the steps involved—especially the precautions. Again, if you're new at this you should practice several times before actually beginning your project.

With reducer fittings you can join pipes of different diameters. Transition fittings allow you to connect pipes of different composition. Remember to always use a dielectric fitting when joining copper pipe to galvanized pipe (page 33).

Joining Flexible Copper Tubing with Flare Joints. *Flare joints* are used on flexible copper tubing when soldering or using a compression joint isn't possible. Because they weaken the end of the pipe, they should be used sparingly. Their use is restricted to exposed plumbing and they are good for hooking up fixtures and appliances. Fairly easy to make, they involve the use of a flare fitting and a special flaring tool. Flare fittings are available in a variety of shapes—Ts, Ys, reducers, and so forth.

Joining Copper Tubing with Compression Joints. *Compression joints* may be made with either rigid or flexible copper tubing. The process is similar to flare joining except that no special tool is needed. Like flare fittings, compression fittings are available in a variety of shapes. A compression joint makes a firm seal; once you have made it, the compression ring cannot be removed.

Disassembling Copper Tubing. Copper tubing is easier to take apart than galvanized pipe. Still, you should anchor pipe securely as any quick jerks could damage joints elsewhere in the run. Where compression fittings, flare fittings, or unions have been used, you can begin by unscrewing them at the joints. Pipes can be cut the same as galvanized pipe, with either a pipe cutter or a fine-tooth hacksaw. Where joints have been sweated and you have the needed space to pull pipes free, use a propane torch for the disassembly. Make sure that water is drained from the pipe before heating it.

Copper Tubing Repair Skills. For leaks in copper tubing that appear at compression or flare joints, you should

USING A PROPANE TORCH

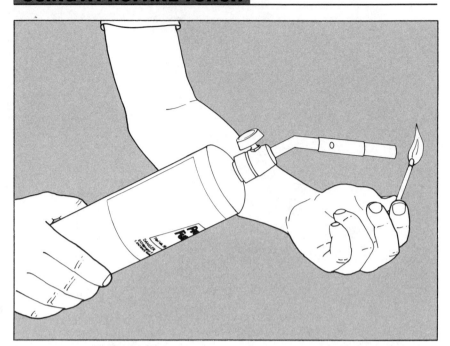

Propane torches used by plumbers consist of two parts: a replaceable metal fuel tank and a combination valve and nozzle assembly. The valve assembly is simply screwed to the threaded fitting at the top of the tank. Light the torch by striking a match, holding it near the nozzle, and slowly turning the valve handle. Continue opening the valve until the flame is large enough for the soldering job. If you make the flame too large, the gas pressure will cause it to extinguish. While working with the torch take care not to tilt the tank too much as the liquid propane will flow upward and block the valve causing the flame to go out.

CAUTION

■ Because the flame of a propane torch is silent and not highly visible it is very easy to forget about. You should always turn it off when you set it aside.

■ Be constantly aware of the direction of the flame.

■ Follow manufacturer's instructions when handling LP gas torches.

■ Keep your work area free of clutter and debris, especially combustibles.

■ A soldered pipe will conduct heat to within 12 inches of the area where the flame is applied. Be careful not to touch pipe anywhere in this area. For extra protection, wear gloves and safety glasses!

■ Keep bystanders well out of the way.

■ Make sure that the pipe you are working on is drained thoroughly. Even a small amount of water may build up steam-pressure and cause a pipe to burst—possibly causing injury.

■ Use a piece of flameproof sheeting or sheet metal in front of flammable surfaces. If a wooden surface becomes charred, wet it with a spray bottle and then check it before leaving the scene of the job.

■ If you are working close to a faucet or valve, remove and disassemble it. Nonmetallic parts such as washers are easily distorted by the heat of a propane torch.

■ Keep a fully charged, all-purpose fire extinguisher at hand.

check the fitting and pipe carefully. Remedy the problem by cutting off the damaged end and making a new flare. For leaks in copper tubing that appear at sweated joints, clean and 'resweat' the joint. When leaks appear in straight runs of copper tubing, you must cut the pipe and add one or two couplings or one or more compression unions.

Support copper pipe in horizontal runs with hangers at every 6 to 8 feet. To prevent galvanic action, use insulated hangers or wrap metal hangers with electrician's tape at any point where the copper pipe touches it.

SOLDERING (SWEATING) COPPER TUBING

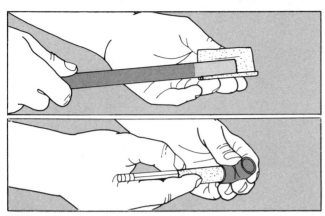

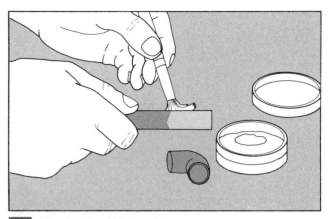

1 After tubing is cut and burrs are removed, both the tubing and the fitting should be cleaned, especially on the inside. Use emery cloth for this purpose and rub until the surface is bright.

2 After cleaning, do not touch the surfaces; any amount of dirt, even a fingerprint, can weaken the joint.
Next apply a thin layer of flux to both of the cleaned surfaces; use a brush to do this.

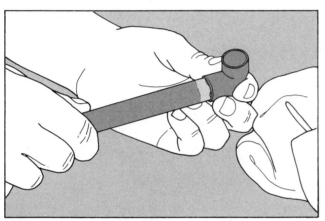

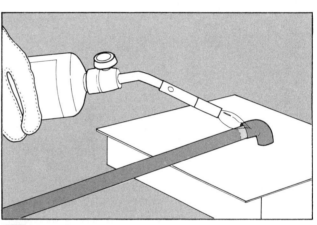

3 Assemble the tubing to the fitting and give it a quick twist to evenly distribute the flux. With a rag, wipe off any flux that has oozed out of the joint. If you are working in a tight space or one exposed to flammable surfaces, first prepare your workspace by mounting flameproof sheeting behind the tubing.

4 Light the propane torch and spread the tip of the flame evenly over both the surface of the fitting and the tubing, all the way around the joint. The surface will be properly heated when the flux bubbles or smokes a little.

5 Test to make sure that the joint has been heated to the correct temperature. Do this by touching solder to both the tubing and the fitting. The solder will melt on contact and be drawn to the joint. Note: If you have heated the joint too much, the flux will burn off and the solder will not flow properly. If the joint cools as you are working the solder into place, the solder will no longer be sucked up into the joint. When this happens you can continue heating the metal surfaces with the torch but you should not let the flame touch the solder itself. (Experienced plumbers know exactly how long to heat a joint; once they begin using the solder, they don't use the torch again.) You will be finished when there is a smooth bead of solder all the way around the joint. Allow this joint to cool for about 10 minutes before going on to the next joint in the run, or, if you're in a hurry, hasten the process by squirting the joint with cold water. The solder will lose its sheen when it is completely hardened.

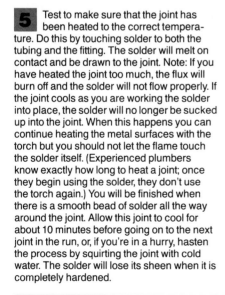

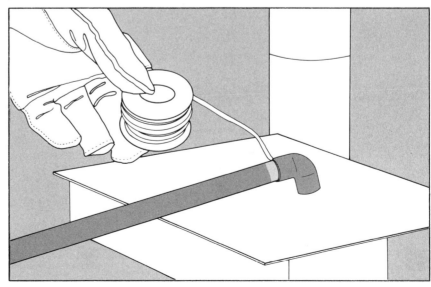

MAKING FLARE JOINTS

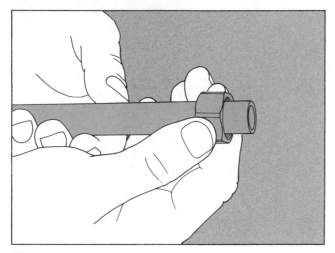

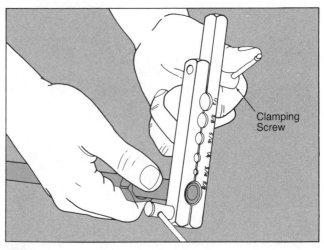

Clamping
Screw

1 Cut and ream tubing as you would for a soldered joint (page 23). Slip the female part of the flare fitting (called a flare nut) over the tubing with its threaded end facing the end to be joined.

2 Place the end of the tubing into the flaring tool at the hole sized for it. The end of the tube should be flush with the die. Tighten the clamping screws.

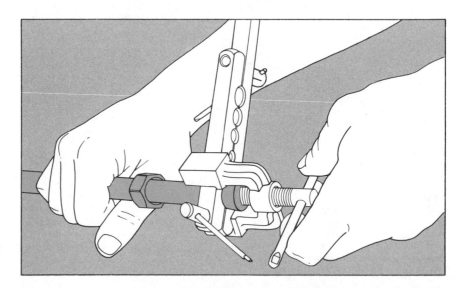

3 Position the ram of the flaring tool into the end of the tube and screw it tightly to make the flare. The end will be flared to an approximate 45° angle. Turn the handle counterclockwise to retract the tool.

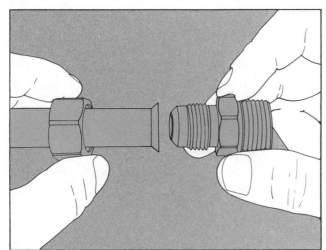

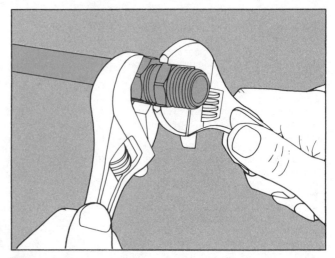

4 Place the domed end of the fitting against the flared end of the tubing, as shown.

5 Screw the flare nut onto the fitting; use two wrenches to tighten the assembly.

MAKING COMPRESSION JOINTS

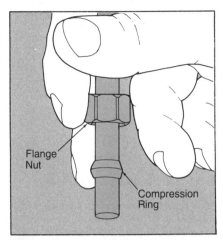

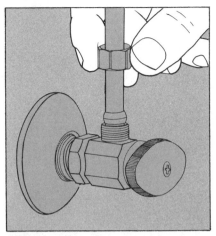

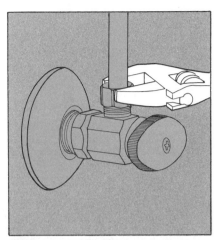

1 Cut and ream tubing as you would for a soldered joint (page 23). Slip the flange nut and then the compression ring onto the tubing.

2 Insert the tubing into the fitting as far as it will go. (In this case the fitting is a shutoff valve.)

CAUTION

Tightening the nut too much will damage the compression ring and keep it from seating properly.

3 Push the compression ring squarely into the joint and then slide the flange nut over the threaded fitting. Screw it down by hand and then give it another one-quarter turn using a smooth-jawed wrench. If the fitting joins two sections of pipe, use two wrenches—one to keep the fitting from turning, and the other to tighten the nut. If the fitting leaks after you turn the water back on, give it another one-quarter turn.

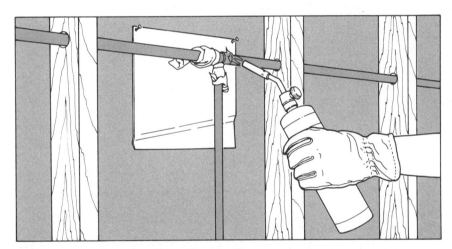

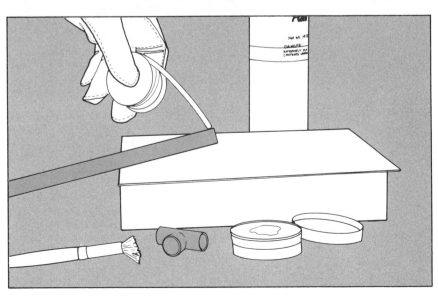

Taking Apart Copper Tubing. Tubing that is joined by compression fittings, flare fittings and unions can all be uncoupled by unscrewing; for soldered joints you can either cut pipe with a fine-toothed hacksaw or, where the ends can be pulled free, use the method shown here. Soft pipe is usually easy to pull free once soldered. Once water is completely drained from the line, begin by bracing the pipe run so as not to strain joints. One way to do this is to wrap pipe at 3-foot intervals with plumber's tape (page 17), tighten the metal tape, and nail it to nearby joists or studs. Next, shield flammable surfaces with flameproof sheeting or sheet metal and wrap wet rags around soldered joints that you don't want to disturb. Melt the joint slowly and evenly with the flame from a propane torch until the tubing can be pulled free.

Stopping Leaks at Copper Joints.
A leak in a sweated joint is usually caused by a faulty soldering job. After draining the line and removing the tubing, clean the mating surfaces of both the fitting and the tubing. Apply flux as you would for a new joint. With a torch, heat the tubing and fitting separately. Cover the fluxed surfaces with solder, allow them to dry and then scour the 'tinned' surfaces with emery cloth. Reassemble the joint and sweat it as described for a new joint.

Compression and flare joints usually leak because of badly cut pipe or poor assembly. Loosen the nut and check to see if the fitting and nut have been cross-threaded. If so, the entire joint should be replaced. In compression fittings, make sure that the pipe end and compression nut are not bent out of shape. Cut off a distorted pipe end or replace a misshapen compression nut. Check the inside of a flared pipe for burrs or gouges. For any of these problems, cut off the damaged end squarely and make a new flare.

THREE WAYS TO REPAIR A LEAK IN COPPER TUBING

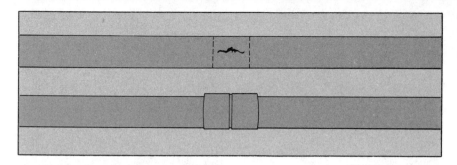

With a Coupling. With a fine-toothed hacksaw, cut out the leaky section of the tubing. Pull the tubing ends together and solder on a coupling.

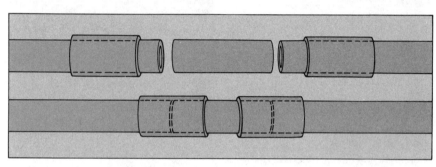

With Two Couplings. If the ends of the tubing cannot be pulled together, cut the tubing apart, cut a piece of new tubing, and solder on two slip couplings.

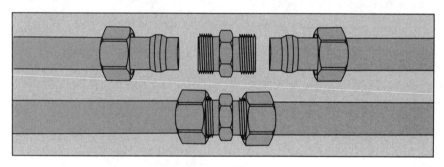

With a Compression Union. If soldering is not desirable, cut the tubing apart and use one or more compression unions to join the tubing.

CUTTING PLASTIC PIPE

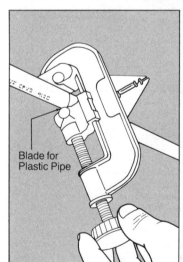

Blade for Plastic Pipe

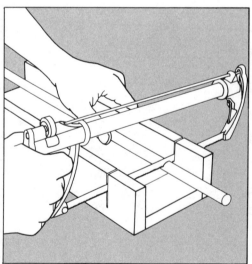

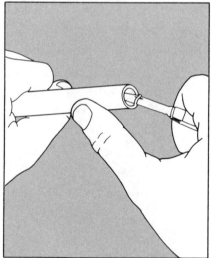

1 Plastic pipe can be cut with a special tube cutter or with a cutter made for flexible copper tubing, as long as you can insert a blade that will work on plastic. Another method, just as easy, is to use a hacksaw (or a backsaw) and a miter box. The hacksaw can be standard (24 teeth per inch).

2 After you have made the cut, remove any rough edges with a pocket knife and finish off with a file or sandpaper. Remove scrapings from pipe, especially inside.

Working with Plastic Pipe

As mentioned on page 15, there are several types of plastic pipe. Though different, they are all handled similarly when they are being cut, joined, and repaired.

Measuring and Cutting Plastic Pipe. Plastic pipe, like copper tubing, is measured by first calculating the length of exposed pipe. Next, add the distances that pipe is inserted within pipe fittings. This will vary according to the fittings being used. Generally, push-in type fittings go in all the way to the beginning of the pipe's shoulder while threaded fittings don't enter quite as far. Pipe is easily cut with a tube cutter or with a hacksaw.

Joining Plastic Pipe. Joining plastic pipe can be tricky because solvent-cement dries very quickly and doesn't always spread evenly. Especially when you're working with fittings that must be positioned correctly, this can require skill. It is best to practice with a few scrap pieces of pipe before attempting to do your actual project. There are many fittings available for joining plastic to other types of pipe.

Plastic Pipe Repair Skills. If plastic pipe is leaking in a straight run, you can repair it by cutting out the damaged section and installing one or two couplings. For leaks at fittings, install two new couplings, two new sections of pipe, and a new fitting. Plastic supply pipe should be supported by a hanger every 6 to 8 feet in a horizontal run. Vertical runs need not be supported. Plastic DWV pipe must be supported every 4 feet or at every fitting (whichever is less) by plumber's tape (page 17) or large wire hangers.

JOINING PLASTIC PIPE

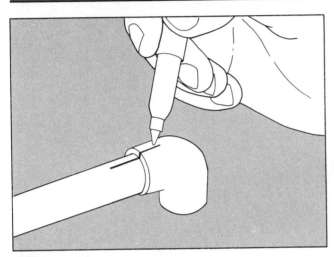

1 Make sure that you are using the correct kind of solvent-cement for the type of pipe that you are joining. Some pipe manufacturers recommend the use of a primer before applying the solvent-cement. With a clean, dry rag wipe off the surfaces of the pipe and the fitting. Test for sizing by pushing the pipe into the fitting; it should go in about halfway. (The joint will leak if it is too tight or too loose.) Try different fittings until you find one that fits correctly and then assemble them exactly as you want them. Score or mark the pipe and fitting, as shown, with guide marks for assembly.

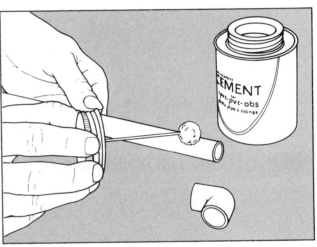

2 Use either fine sandpaper or a specialty pipe cleaner to remove the gloss and any oil or moisture from the surfaces of the joint. Next, with the brush that comes with the solvent-cement, spread a light coat of solvent-cement on the outside of the pipe and inside of the fitting. Quickly spread another light coating on the pipe.

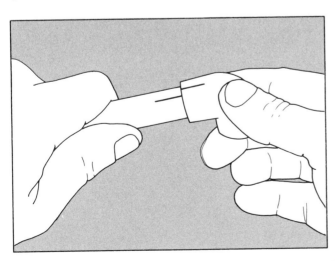

3 Immediately after spreading on the solvent-cement, insert the pipe into the fitting close to your positioning mark. Then give the pipe a quarter turn to align the two marks. There should be an even bead of solvent-cement all the way around the joint. If there is not, quickly pull the pipe out and apply one more coat of solvent-cement. When the joint is set correctly, hold it in place for 30 seconds or according to the manufacturer's directions. Wait at least 3 minutes before starting to work on the next joint. Wait at least one hour or, preferably, overnight before running water through the new piping.

TWO WAYS TO REPAIR A LEAK IN PLASTIC PIPE

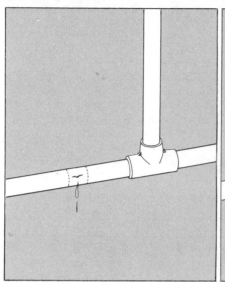

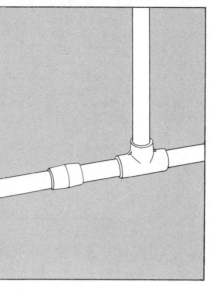

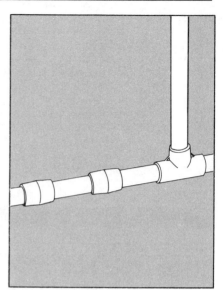

With a Coupling. With a hacksaw, cut out the damaged section of the piping. After cleaning, pull the ends together and, with the instructions given on page 29, join the pipe with a coupling.

With Two Couplings. If the ends of the pipe cannot be pulled together, cut the pipe apart, cut a piece of new pipe and join it with two couplings.

STOPPING LEAKS AT PLASTIC JOINTS

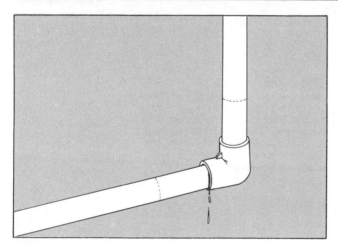

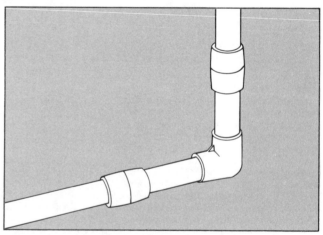

1 When you have a leak in plastic pipe at a fitting you can try to fill the joint with solvent-cement but this rarely works. A better method is to cut off pipe on each side of the leaky joint and replace it with new piping.

2 Use two new couplings, two new pieces of pipe, and a new fitting to make the repair. Work from one end to the other, as described on page 29.

Working with Cast Iron Pipe

If your local plumbing code permits it, use plastic pipe instead of cast iron wherever possible. If you can't use plastic, then be sure to use hubless cast iron pipe for any installation or repair work; it is much easier to work with than the old-style hub type. Also, for more information about cast iron pipe, see pages 42-43 regarding unclogging drains.

Measuring and Cutting Cast Iron Pipe. When you're replacing pipe, simply measure the length of the pipe that you're replacing. Accuracy is important since cutting this pipe is quite a chore and the pipe is expensive. Cut pipe with a rented pipe cutter or with a hacksaw and chisel. If the pipe is vertical, brace it by using plumber's tape nailed at intervals to framing members of the house.

Joining and Supporting Cast Iron Pipe. Cast iron pipe is easily joined with special neoprene gaskets and stainless steel bands. It should be supported with a sturdy hanger in a horizontal run at every fitting and every 5 feet and, in a vertical run, at every 5 feet. (See page 32.)

CUTTING CAST IRON PIPE

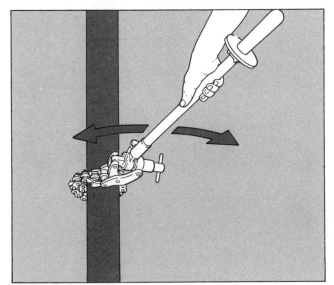

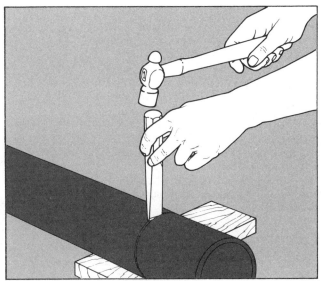

With a Pipe Cutter. The easiest way to cut cast iron pipe, and the only way if you have limited access space, is to use a special tool called a *soil pipe cutter.* These are usually available at rental stores. First, secure a vertical pipe above and below where the cut is to be made, using plumber's tape and attaching it to nearby studs or joists every 2 feet or so. Mark the pipe to be cut and then wrap the chain-like section around it placing the cutting wheels on your mark. Secure this section to the body of the tool, tighten the knob and work the handle back and forth, gradually tightening as the cut deepens. The pipe will eventually snap.

With a Hacksaw and Chisel. If you are working with new cast iron pipe, you can cut it with a hacksaw. Begin by marking the pipe on its entire circumference and raising it up on a block. With the hacksaw, saw evenly until there is a 1/16-inch groove. With a hammer and 3/4-inch cold chisel, tap the groove all the way around the pipe. It might take three or four revolutions before the pipe snaps.

JOINING HUBLESS CAST IRON PIPE

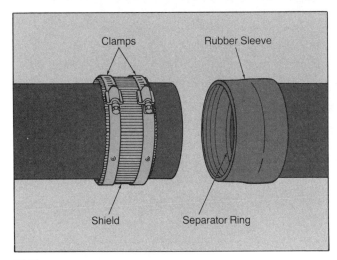

Clamps   Rubber Sleeve

Shield   Separator Ring

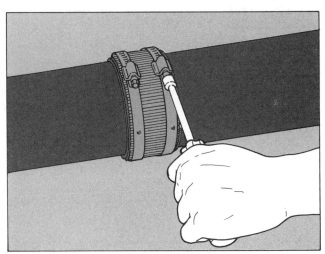

1 To join hubless pipe you'll need a special sleeve and clamped shield; these are available in sets to match the size pipe that you are working with. First, slip the rubber sleeve onto the end of one pipe and make sure that the end of the pipe is snug against the sleeve's separator ring. Next, slip the steel shield over the end of the other pipe. Make sure that the clamps of the shield will be accessible for tightening.

2 Push the end of the pipe that has the shield into the rubber sleeve as far as it will go—until it butts against the separator ring. Position the shield over the rubber sleeve until it is covered completely. Tighten the clamps with a screwdriver or a socket wrench.

Working with More Than One Kind of Pipe

The evolution of new plumbing materials has widened the scope for the home plumber. Depending on your code, there might be an opportunity to make replacements or additions with easy-to-use materials—pipe unlike what you already have in your home. Luckily, there are transition fittings that make this possible. With the skills you have gained, you can join copper pipe to galvanized pipe, or plastic pipe to copper or galvanized. To prevent galvanic action (page 8), however, be sure to use a dielectric fitting between unlike metals (copper and galvanized).

Calculating Pipe Dimensions

The chart shown below will aid you in determining pipe dimensions. If you know the inside dimension of a pipe (ID), then you can easily determine its outside dimension (OD), and vice-versa. Also, a given size fitting such as a ¾-inch copper ell will have a depth of ¾-inch from end to shoulder. Compare this to the ¾-inch plastic fitting which has a longer ⅝-inch depth. Use this chart for measuring pipe and for ordering supplies. Most importantly, always choose replacement pipe with the same *inside diameter* as the pipe you are replacing.

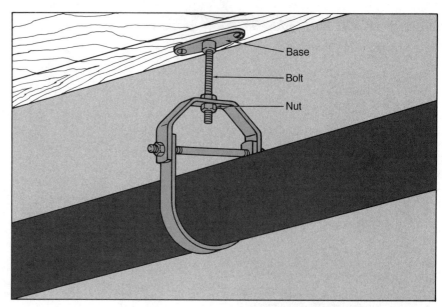

Supporting Cast Iron Pipe. Hubless cast iron pipe, because of its weight and because of its somewhat flexible joints, should be supported in a horizontal run at every fitting and/or at 5-foot intervals. For this purpose, use a special clamp (shown here). Attach it by first screwing the base into the ceiling. Then assemble the clamp around the pipe and attach it to the base with the bolt and nut. Adjust the tension by tightening the nut. In a vertical run, support pipe at 5-foot intervals. Use either plumber's tape (page 17), stack clamps (page 79), or a special hanger similar to the one shown here.

PIPE DIMENSIONS (In Inches)

PIPE COMPOSITION	INSIDE DIAMETER (ID)	OUTSIDE DIAMETER (OD)	DEPTH OF FITTINGS	PIPE COMPOSITION	INSIDE DIAMETER (ID)	OUTSIDE DIAMETER (OD)	DEPTH OF FITTINGS
Galvanized	⅛	⅜	¼	Plastic	½	⅞	½
	¼	½	⅜		¾	1⅛	⅝
	⅜	⅝	⅜		1	1⅜	¾
	½	¾	½		1¼	1⅝	¹¹⁄₁₆
	¾	1	⁹⁄₁₆		1½	1⅞	¹¹⁄₁₆
	1	1¼	¹¹⁄₁₆		2	2⅜	¾
	1¼	1½	¹¹⁄₁₆		3	3⅜	1½
	1½	1¾	¹¹⁄₁₆		4	4⅜	1¾
	2	2¼	¾				
Copper	¼	⅜	⁵⁄₁₆	Cast Iron	2	2¼	2½
	⅜	½	⅜		3	3¼	2¾
	½	⅝	½		4	4¼	3
	¾	⅞	¾		5	5¼	3
	1	1⅛	¹⁵⁄₁₆		6	6¼	3
	1¼	1⅜	1				
	1½	1⅝	1⅛				

JOINING DIFFERENT KINDS OF PIPE

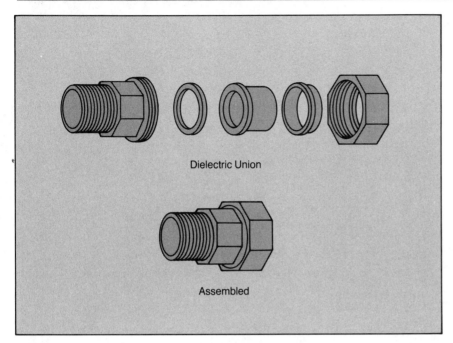

Dielectric Union

Assembled

Joining Copper Tubing to Galvanized Pipe. Special fittings are needed when connecting one kind of pipe to another. To prevent galvanic action from occurring when joining copper with galvanized pipe (page 8), you should use a dielectric union which contains special insulators for this purpose. Caution should be used, however, to make sure that the pipes you are joining with this union are not part of the electrical grounding system (page 8).

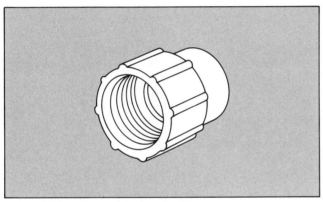

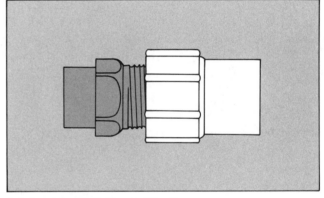

Joining Galvanized Pipe to Plastic Pipe. Use a special galvanized iron-to-plastic adapter. Made of plastic, one end has interior threads for the insertion of iron pipe; this joint should be reinforced with fluorocarbon tape. The other end is also 'female' so that plastic pipe is simply fitted into its end and joined with solvent-cement. When you purchase a plastic adapter, specify whether it will be used in a hot or cold supply line.

Joining Copper Tubing to Plastic Pipe. Use two adapters for this process. The one on the left, made of copper, is sweated to the copper pipe. The kind shown here is 'female' with a collar inside it for accepting a smaller-diameter pipe. Once this adapter is in place, the other, threaded end of it is joined to the 'female' plastic adapter and reinforced with fluorocarbon tape. Plastic pipe is then inserted into its opposite end and joined with solvent-cement. When you purchase adapters, specify whether they will be used in a hot or cold supply line.

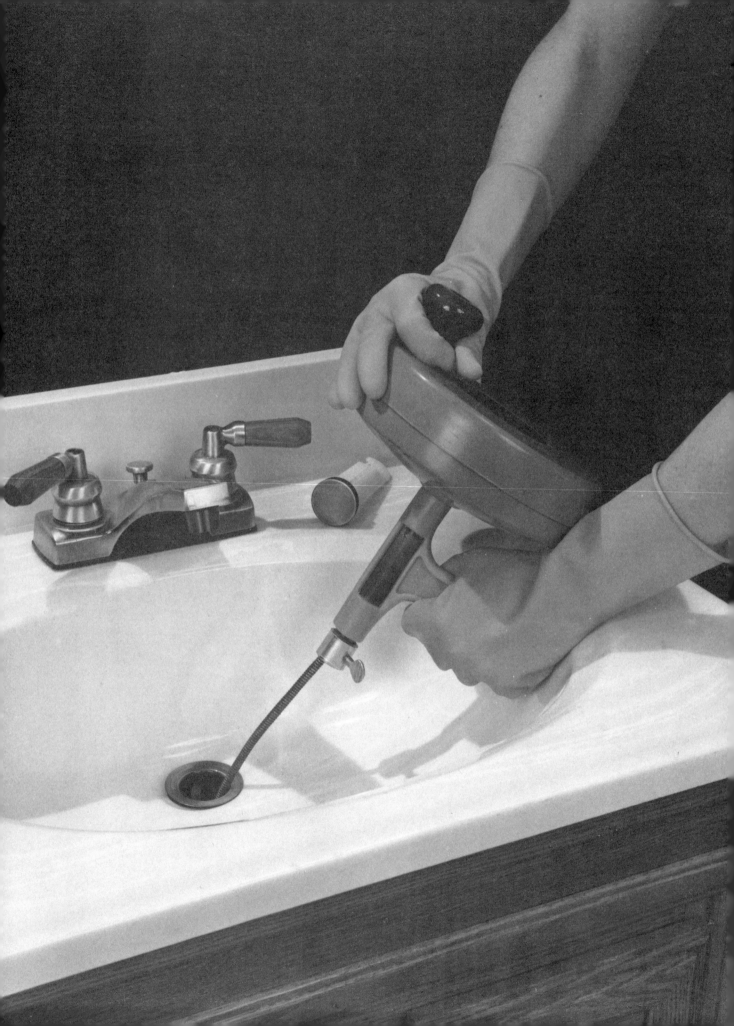

Common Plumbing Repairs

T his chapter and the one that follows are notable because it's here that you'll learn what's involved in repairing most of your plumbing fixtures. (Chapter 5 is about faucets but most other common repairs are covered in this chapter.) By doing simple repair jobs yourself you will save a great deal of money. But even in the cases of larger, more complicated repairs that warrant a call to a plumber, by reading the repair steps, you'll know what it is that you're paying for.

You'll learn how to deal with plumbing emergencies. Often, people panic because they aren't sure of how to handle leaks, clogs, or frozen pipes. At the other end of the spectrum are repair projects for long-overlooked problems such as a faulty sink stopper. This is the kind of repair that gets put off for years because it's not quite worth the price of a plumber. Be sure to follow the guide for non-emergency repairs; these tips will help you avoid unnecessary shopping trips and other similar dilemmas.

Practically everything is covered— from very tiny pipe leaks to the big main cleanout in your basement. There's also a toilet trouble-shooting chart so you can make a quick analysis of what's involved in repairing this sometimes puzzling apparatus.

Repairing your own plumbing can actually be fun. You can start making small repairs one-by-one in your spare time and then set aside the larger projects for weekends. Eventually, you'll have your system in top-notch condition! Plus you'll develop a confidence about what to do in case of an emergency. As always, use safety and adhere to the cautions.

The Plumbing Crisis ... and How to Respond

Though your home plumbing is, in all probability, composed of sound and durable materials, there is, as with all things, a time when it simply wears out. Such is the case with leaky pipes that have long served their purpose, and with the parts of faucets and other fixtures that get continuous use in a household. Sometimes a tiny drip will be bothersome but its repair can be put off until a rainy day. Other times, during a plumbing emergency, the rainy day (or flood!) is precipitated by the problem.

Then you need to act—and act fast.

Obviously, the best way to handle an emergency is to be prepared for it. The first thing to do is to educate yourself about your plumbing system. Locate the main shutoff valve (page 9). Check to see which fixtures have shutoff valves of their own and operate them to make sure that they work properly. Check any other shutoff valves in your home; if you think that you won't remember what they control, mark them.

Have plumbing tools handy. If you haven't purchased all of the tools on page 12, then at least have the necessary ones for emergencies, such as a plunger and wrenches. Supplement this with an emergency plumbing kit—a kind of 'first aid kit for pipes'. Put all these materials together in a box and keep everything organized so that you can locate them easily.

When to Call for Help. The most common emergency situations are burst or leaking pipes, frozen burst or leaking pipes, and clogs in fixture drain-pipes or main drainpipes. The wisest thing to do is to first assess the situation and then shut down the water. In general, the smaller jobs can usually be remedied by the homeowner and you can, of course, save yourself a lot of money by doing the repair yourself. If your emergency happens late at night or on a weekend, you'll pay top price when calling a 24-hour serviceman; so, if at all possible, wait until regular working hours to make your call. Then, call several places for estimates and question each business about their procedures, like cleanup, and their policies.

There are times when the cost of a professional is more than worth it. If a flood is doing permanent damage to your house, destroying ceilings and leaking or rising around wiring or fixtures—call the fire department and your electric utility company. Do not wade through water or touch electrical equipment while standing in it. If you've worked with a plunger to unclog a toilet and you find the next, fouler, step of using a snake too offensive—by all means call a plumber. At times, also, plumbers will have industrial-size equipment, enabling them to get the job done much faster.

Emergency Repairs for Pipes. Since supply pipes operate under pressure, they are the most likely source of leaks. If you have the time to, repair pipes permanently by using the methods in the skills chapter (pages 18-33). But for pipes that spring leaks when you're 'off duty' or otherwise detained, make a temporary patch in one of the following ways: Use epoxy, tape, rubber hose, clamps, or even a pencil tip to plug up a hole.

Your choice of materials will be determined by the size and location of the leak. For any of these to work, you must first drain the pipe and dry the surface. (If the leaky pipe is frozen, on the other hand, go ahead and make repairs before completely thawing it; try to keep it as dry as possible.) Clamps

work well in straight pipes, whereas epoxy is useful at joints where clamps will not fit. A pencil tip or tape usually suffices for smaller holes. These repairs will be good for a few weeks to a few months—enough time for you to arrange for a permanent repair.

Leaks—Unwanted or Wasted Water. Leaks are usually obvious, either by sound or by sight, and when they appear you should take special measures to minimize their damage. Everybody knows the bucket trick associated with roofs and rainwater and truly, buckets, basins, and mops will be useful. But when you see supply water dripping from your ceiling the most important step to take is to shut off the nearest supply valve. Use dropcloths under any 'catching' utensils, especially if the drips are substantial. If flooding begins, use towels or rugs to create 'dams'. When it gets out of hand, call your fire department.

Most ceiling leaks are easy to locate, being directly above the major drips. Occasionally, however, water will run a short distance along a joist and then stain a ceiling at another point. Leaks running down a wall indicate pipe trouble in a vertical run; these can be more difficult to locate, sometimes necessitating the removal of the wall surface.

Sometimes leaks are elusive and the real emergency is your sky-high water bill. If this is the case, begin your search for a leak by shutting off all faucets and valves and then inspecting the water meter. If the dial moves, water is being lost somewhere in the system. With a flashlight, continue the search in the basement or crawlspace of the house. More than likely the leak will be found at this lower level since leaks elsewhere

in the home provide visible and/or audible evidence.

All About Frozen Pipes. Most homes in cold weather climates have plumbing installations that resist this problem. But good home construction will prove inadequate in case of an emergency such as when there is an unusually harsh and long cold spell or a power failure that shuts off the home's heat. If and when this happens, here are some preliminary steps, both electrical and nonelectrical, that you should take to prevent pipes from freezing:

■ Keep a very small trickle of water running from your faucets.

■ Open the inside doors of your house to distribute heat evenly.

■ Keep light bulbs burning near exposed pipes. (Watch out for nearby combustibles, though.)

■ Place a small heater in your basement.

■ Wrap pipes with foam, tape, newspapers, or heating wires.

■ If you are vacating the home, completely shut down the system (page 124), including the electricity.

The first sign that your pipes are frozen is usually a faucet that refuses to yield water. When you make such a discovery, you should first turn off your water at the main shutoff valve (page 9). Next, make sure that the 'frozen' faucet is open. If you spot obvious leaks, make emergency repairs like the ones on page 37.

If no leaks are evident, prepare the area for large surprise leaks. Use dropcloths and have buckets and pails ready. Then use one or more of the methods shown here to thaw pipes. Be sure to follow the safety precautions, especially if you decide to use a propane torch.

PLUMBING KIT FOR EMERGENCIES

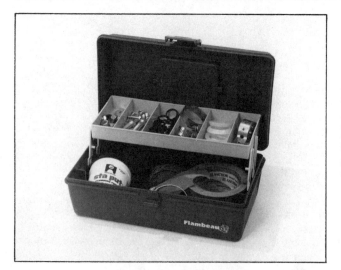

Assemble an emergency plumbing kit by purchasing these items at a hardware or plumbing supply store: several nuts, bolts, and washers of various sizes; assorted washers and screws for faucets (often prepackaged as kits); O-rings; pipe joint compound and a roll of duct tape. Supplement your kit with these common household items: a length of old garden or radiator hose; scraps of old rubber (from rubber gloves, a water bottle, an inner tube, or the like); a few wire coat hangers; pieces of sheet metal, such as from a coffee can and a few automotive hose clamps.

PATCHING TINY LEAKS

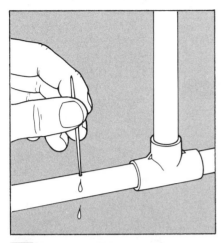

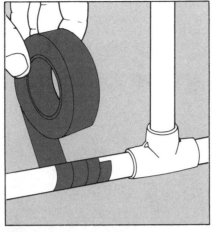

1 To stop a tiny leak in a supply pipe, simply push a toothpick into the hole and break it off. A soft pencil lead also works well.

2 Dry off the pipe surface completely and then wrap the plug with duct or electrical tape. Wrap it with several turns to the right and left of the plug.

CAUTION
To patch a leaking pipe, begin by shutting off the water at the main shutoff valve (page 9) and draining the pipe at the nearest fixture outlet.

PATCHING A LEAK WITH RUBBER HOSE

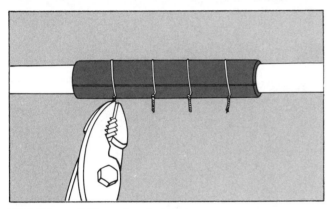

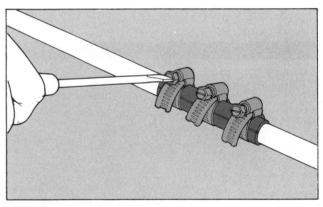

With Wire. Split a section of rubber hose lengthwise to fit around the pipe. Wrap strong but flexible wire around the hose at 1-inch intervals and twist together tightly with pliers.

With Hose Clamps. Split a section of rubber hose lengthwise to fit around the pipe. With automotive hose clamps, secure the hose in at least three places, tightening the clamps as much as possible.

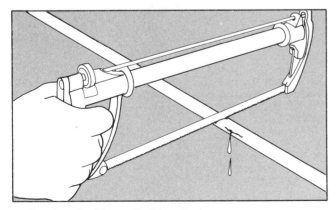

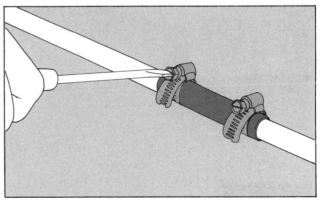

By Cutting the Pipe. For this method you will have to turn the water off and then cut out the section of pipe where the leak is. Cut a piece of rubber hose with a slightly larger diameter than the pipe. Cut it a bit longer than the pipe.

Use hose clamps to secure the hose into place; one at each end should suffice.

Begin the thawing process close to the open faucet and gradually work toward the frozen area. Vapor and water will escape from the faucet. If you are confident that there are no leaks in the pipe or you have already repaired them, you can turn on the main shutoff, but only slightly. This extra bit of pressure will help to push slush and ice a little further through the line.

Dealing with Clogged Drains. In spite of our good intentions, we all sometimes goof when it comes to drainpipes. Garbage, gunk, and hair get forced down these hard-working elements of the plumbing system. Toddlers delight in watching tons of toilet paper dissolve in the toilet bowl. These and many other situations cause clogs and when the drainpipes 'cough', nobody feels worse than the person left to do the unclogging.

Clogged Toilets. If a toilet flushes incorrectly or you suspect that it's clogged—do not flush it again! On the other hand, if you see that an overflow is about to occur, very quickly remove the tank lid, reach inside and push the stopper into the valve seat (page 51).

Toilets are usually clogged in their traps and the simple use of a plunger will push the clog through. Try several times with a plunger before resorting to the use of a snake or closet auger. These tools are made especially for toilets and

their cost is nominal compared to hiring a plumber. Working the snake will be messy, but it usually does the trick. If the snake fails, then the next step is to check the main drainpipe system (page 42).

A WORD ABOUT CHEMICALS

The best way to avoid a clogged drain is to treat it when it's sluggish. No doubt you've heard the commercial praises of chemical drain cleaners, and they do help to keep some drains clear if used on a regular basis. But they also have their drawbacks.

Chemical drain cleaners should never be used with a septic tank system. They are caustic and, depending on their composition, will harm pipes and porcelain. Most importantly, they should not be used when pipes are already clogged; they will only worsen the situation. Once you have used them, you cannot use a snake or plunger safely. Still, if you

insist on using such products, be sure to follow the manufacturer's instructions precisely. Wear eye protection and gloves. Also protect exposed skin.

There are alternatives to using chemicals for drainpipe maintenance. Among them are cleaning out floor drain strainers, cleaning pop-up stoppers regularly and running a garden hose into your DWV pipe on the roof. This last procedure can be done on a seasonal basis when you're cleaning out your gutters and downspouts. Run water full-force into all vents for about two minutes.

METHODS FOR THAWING FROZEN PIPES

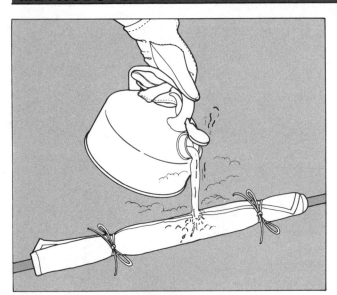

With Hot Water. If a pipe is horizontal and easily accessible, use this method. Simply tie absorbent rags around the frozen section of piping. Pour hot water in a steady stream over the rags until the ice thaws and water in the pipe runs freely.

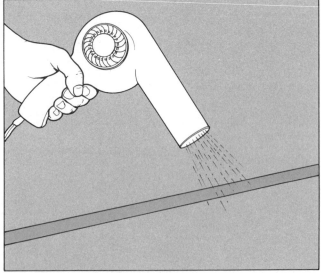

With a Hair Dryer. Hair dryers can be used to thaw pipe although you should make sure that you are using a grounded appliance plugged into an outlet with a GFCI (Ground Fault Circuit Interrupter). Also check the dryer's body and plug for damage. Work the blowing dryer back and forth and test periodically for running water. (This method takes patience so be prepared to wait awhile.)

WARNING

Overheating a pipe can be very dangerous. Water inside the pipe could begin to boil and steam could form, resulting in a dangerous explosion. Therefore, open a faucet before starting.

CAUTION

To thaw a frozen pipe, begin by shutting off the water at the main shutoff valve (page 9) and draining the pipe at the nearest fixture outlet.

UNCLOGGING TOILETS

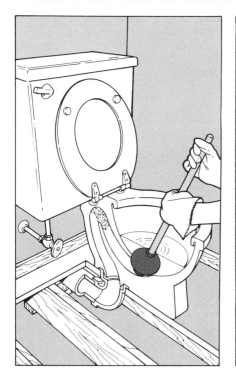

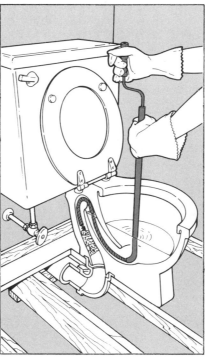

(L) With a Plunger. Use a plunger with a conical base for this job. With the normal amount of water in the bowl, push down rapidly 10 or 12 times, or until the toilet appears to be draining. If you think you have pushed the stoppage through, pour in a bucket of water to make sure. If this method is unsuccessful, use a snake.

(R) With a Snake. This type of snake or auger is called a *closet auger* because it is specifically designed for toilets. With the normal amount of water in the bowl, push the snake into the trap until it stops. Turn the crank handle clockwise and alternately tighten the thumbscrew on the handle. When you feel blockage, jiggle the long base of the snake slowly while continuing to crank. When you withdraw the snake, be sure to have a bucket ready to put it in, as it will pull with it much of the gunk from the clog. This process should in time clear out any blockage in a toilet trap. If it does not, you should call a professional plumber to diagnose the problem and fix it with special equipment.

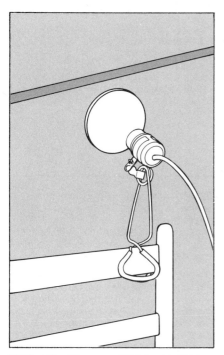

With a Heat Lamp. If the frozen pipe is in a hard-to-reach area, a heat lamp can be useful for thawing. Insert the bulb into a portable work lamp and clamp it to a sturdy ladder, chair, or fixture near the pipe. Be careful not to place the lamp closer than 6 inches to painted or wallpapered walls or other combustibles—it will scorch or set fire to such surfaces.

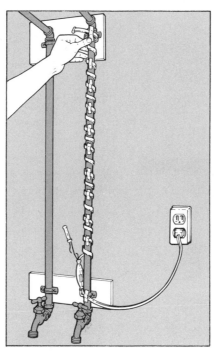

With Heating Tape. Heating tape is not only good for emergency situations, it can also be used as a preventative measure; choose a tape with a thermostat for this purpose. Follow the manufacturer's instructions which will usually include the following steps. Wrap heating tape around the pipe, approximately six turns per foot. Secure masking tape to the pipe at every other turn. Spiral the tape up the frozen section of pipe, allow it to heat and check periodically for thawing.

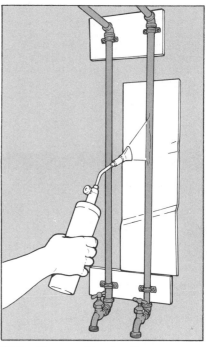

With a Propane Torch. Propane torches are aids for thawing pipe when there is a power failure. They work quickly but require a special attachment, a *flame spreader,* and they must be used with caution. First, make a backing with metal sheeting for your working area. Then, with the flame spreader attached, move the lighted torch back and forth on the piping to warm it. Use your free hand to touch the pipe and check it for overheating; it should not be too hot to touch.

Clogged Sinks, Tubs, and Showers.
Unclogging a sink or lavatory isn't difficult and it shouldn't be a dirty job if you
follow the instructions carefully. Begin
by using a plunger. If that doesn't work,
try a snake. With the snake you will use
a process of elimination to locate the
clog. First work the snake at the drain,
then at the cleanout, and finally, beyond
the trap at the drainpipe.

Clogged tub drains will not respond
to plunging, so use a snake on them. If
the tub has a P trap, you should work at
the overflow pipe; if it has a drum trap
you'll have to work at the trap, snaking
first in the direction of the drain, then in
the opposite direction.

Shower drains are rarely cleared
with a plunger so it's best to begin with
a snake. If this fails, use the force of

water pressure to break up the blockage. Usually effective, this involves running a garden hose to the shower and
working with a helper.

When your efforts at unclogging
fixture drainpipes fail, the next step is
to explore the main drain system for
problems.

UNCLOGGING A TUB

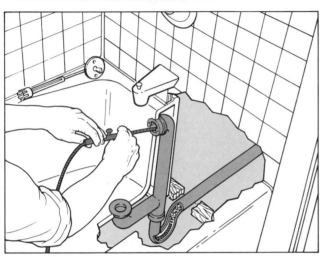

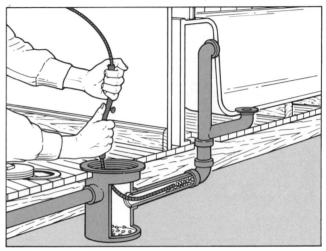

A Tub with a P Trap.   Begin by removing
the overflow plate and pulling out the pop-up
assembly. Feed the snake into the overflow
pipe and push and probe as described on
page 39. This will probably remove the
blockage; if it doesn't you must remove the
trap or, if it has one, the cleanout plug, and
use the snake on the main drain.

A Tub with a Drum Trap.   Older homes
sometimes have tubs with drum traps which
are adjacent to the fixture. Use an adjustable
wrench to slowly remove the drum trap cover
and gasket; have plenty of rags ready to mop
up trapped water. Scoop out any debris from
the trap and then check with the snake for
further clogging toward the tub. Use the snake
to remove the blockage, first in the direction
of the tub and then toward the main drain.

UNCLOGGING A SHOWER DRAIN

(L) **With a Snake.**   Simply remove the
strainer over the drain opening and insert the
snake, pushing until you hit the clog. Work
the snake as described on page 39, until the
drain is clear. If this method doesn't work, use
a garden hose.

(R) **With a Garden Hose.**   For a deep
clog in a shower drain use a garden hose.
Attach one end to a nearby faucet using a
threaded adapter. Push the other end into the
drain trap and then pack rags solidly around
the drain. Turn the water on full force and
then quickly off, all the while holding the rags
in place. (It is best to use a helper for this
step.) The high water pressure should force
the blockage and clear the pipe.

WARNING

When finished, promptly remove the
water hose from the drain. Leaving it in
exposes it to a backup of sewage which
could occur if there were a sudden drop
in water pressure.

UNCLOGGING SINKS

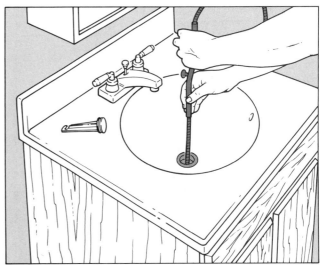

With a Plunger. Several steps must be taken in order to do this correctly. The sink should be approximately half-full of water with at least enough to cover the plunger cup by several inches. The plunger cup should cover the drain opening completely and the rim of the cup should be covered with petroleum jelly to strengthen the seal. Plug up the overflow vent of the sink with wet rags. Once you have taken all of these steps, begin the process. Lower the plunger into the sink on an angle to avoid getting air trapped under it and then push down rapidly while keeping the plunger upright. Do this 15 or 20 times in a row and repeat two or three times before quitting. If the plunger does not force the blockage, try using a snake.

With a Snake. First pull out the pop-up stopper, if there is one, or remove the sink strainer. Next, insert the snake into the drain until it reaches the clog and cannot go further. Use the snake in the same way as desribed for unclogging toilets; continue until it is cleared. If this fails to work, try using the snake through the cleanout.

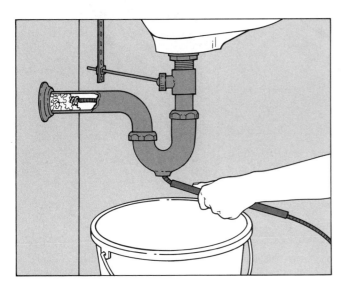

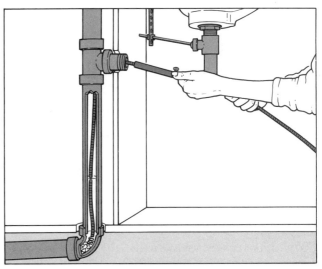

Through a Cleanout with a Snake. The cleanout is an opening at the bottom of the trap that provides easy access to a clog (not all traps have them). With a bucket underneath the trap, remove the cleanout plug. Then, insert the snake and work toward the clog either up toward the drain or further along in the pipe, another possible spot for the clog. If you have cleaned out the trap and cleared the blockage beyond the cleanout and still have a clogged drain, then you must proceed to unclogging the drainpipe.

Through a Drainpipe with a Snake. For this procedure, you need to remove the trap (page 46) using a bucket to catch any spill. Insert the snake into the drainpipe which will likely be in the wall; push it in until it will go no further. Use the snake as described on page 39 for unclogging toilets. If you can't reach the clog with the snake then you probably have a main drain clog (page 42) which will have to be tackled at the main cleanout, soil stack or house trap.

Clogs in Main Drains. There are some clues to main drain clogs. If more than one fixture is clogged or drains sluggishly, then the problem is likely to be here. Also, if there are noticeable odors throughout the house, then the main vent stack is more than likely clogged. One way to take care of this latter problem is to work from the roof with a snake; but because this requires equipment usually beyond the scope of the homeowner and because it is very dangerous, this method is not detailed.

The methods outlined include unclogging at the cleanout and at the house trap. Both of these involve working with raw sewage. They don't require special skills, just the strength to open the plugs and the willingness to work in a mess. As a last resort (or maybe as a first one!), call a plumber or professional drain-cleaning company. Because they come armed with heavy-duty, high-powered augers, they can usually solve your problem very quickly and efficiently.

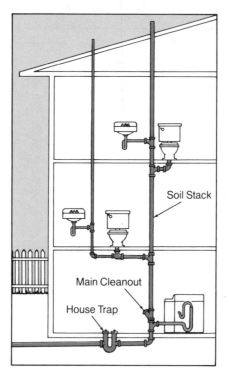

Locating the Cleanout and House Trap. Shown here is the main drainpipe system. Note the location of the main cleanout—under the soil stack in the basement and the house trap, located beneath the floor. When your efforts to unclog drains at the fixture fail, you need to try unclogging in this system.

Soil Stack

Main Cleanout

House Trap

UNCLOGGING THE MAIN DRAIN AT THE CLEANOUT

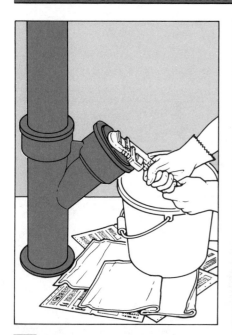

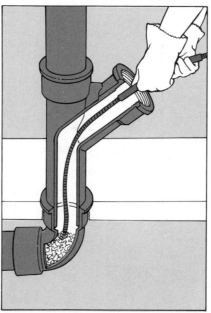

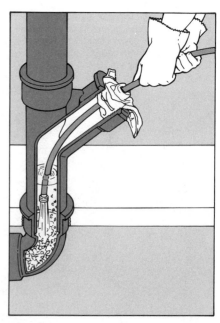

1 Open the cleanout first. Be ready to collect waste in a large pail and have plenty of newspaper spread out in your work area. Remove the plug with a pipe wrench, working very slowly in case flooding occurs.

2 Use a snake to clear out the blockage. The snake should be long enough to reach the sewer outlet, in case the clog extends that far. After you have cleaned it out, flush it with water and then replace the plug after coating it with pipe joint compound.

An optional method is to use a garden hose. Follow the same directions and precautions used for unclogging a shower drain (page 40).

WARNING

Sewer gas has been known to contain noxious as well as flammable gases. Cleanouts should not be left open nor should traps be left empty of water for any undue length of time. If you do detect sewer gas, be sure to provide adequate ventilation.

Defective Sink Strainers

The sink's drainage system begins with the strainer which opens and closes the drain and also traps large particles. You can suspect that there is 'strain' on a strainer when the sink fails to hold water or when leaks appear under the sink.

There are two types of strainers, one containing a locknut, the other held in place by a retainer. They are very similar in design; the locknut type is most common and is always used with a stainless steel sink. Both types contain a rubber washer and metal gasket and, often, a simple replacement of these parts is all that is required (page 44).

Corrosion of the strainer body might also be evident, and, if so, you should consider total replacement. Strainers aren't particularly expensive and a new one will enhance the appearance of the sink.

Anatomy of a Locknut Strainer.
There are two types of sink strainers—the locknut type (shown here) and the retainer type. The locknut is the most common and is generally used on stainless steel sinks. The retainer type, although not shown, differs in that in place of the locknut there is a plastic retainer held in place by three screws under the sink.

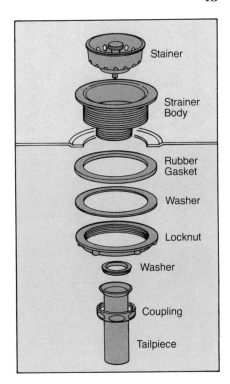

Stainer

Strainer Body

Rubber Gasket

Washer

Locknut

Washer

Coupling

Tailpiece

UNCLOGGING THE MAIN DRAIN AT THE HOUSE TRAP

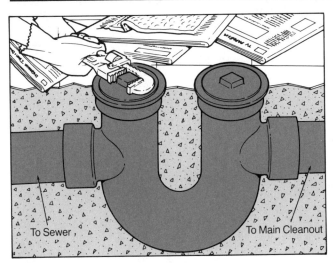

To Sewer

To Main Cleanout

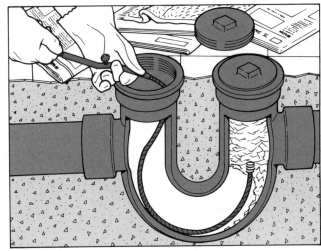

1 The house trap is identified by the two adjacent cleanout plugs; these are usually found at floor level if the main drain runs under the floor. First determine which direction the drain flows toward the sewer; the plug closest to the sewer is the one that you will open. This job can be quite messy so prepare the area with thick stacks of newspaper and rags. Open the plug with a pipe wrench, very slowly in case of flooding. If no water leaks out as you remove the plug, then the clog is located either in the trap or somewhere in the line between the trap and the main cleanout. In this case, proceed to Step 2. If water seeps out as you remove the plug, the clog is beyond the house trap. You can either try removing it yourself, which could be a long and messy job, or hire a plumber.

2 With the plug completely removed, slowly feed the snake into the trap toward the main cleanout. When you meet resistance, do not work the snake the way you normally would—trying to break up the blockage all at once. Instead, probe slowly as if to make a small hole; this allows the water to seep out and prevents gushing water and waste from spilling all over the floor. When water no longer seeps out, remove the other house trap plug and clean the entire trap out with a wire brush.

When the clog is in the pipe between the house trap and the main cleanout, push the snake through the second trap toward the main cleanout; work with the snake slowly, as described above. When finished, recap both cleanouts and then flush the piping with a garden hose, beginning further upstream or at the main cleanout.

A Non-Emergency Repair Guide

Some plumbing problems, like leaky faucets and running toilets, are simply ignored by the homeowner and saved as Saturday repair projects. Unlike emergency situations, these jobs afford you the time you need to work with control and accuracy. However, since many of the projects involve fixtures, they are liable to be tricky. Use the following guidelines to make your work flow as smoothly as possible:

■ Use patience; work slowly and methodically.

■ Don't ever use great force on a part that is corroded and 'frozen'. Instead, apply penetrating oil and wait, overnight if necessary, to remove it.

■ When dismantling a fixture, line up all of the parts, one by one, exactly as they faced each other. This will be a tremendous help during reassembly.

■ Guard against the loss of parts and damage to fixtures. When working at sinks, plug up drains and line the basins with towels. A small dropped part or a large dropped tool can create additional problems in your repair procedure.

■ If a part is corroded or worn, even if it is not causing the problem, replace it. Doing it now is much more sensible than having to go through the disassembly again in the near future.

■ Purchase replacement parts at a plumbing supply store. The selection will be much more complete and the salespeople more knowledgeable than at a hardware store.

■ Buy fixture parts made by the same manufacturer as your fixture; often these will be the only parts that will work.

■ If the search for parts becomes futile, as often happens with long-lasting devices, consider changing the entire fixture. Installation will be easy, the new appearance should prove satisfying, and repairs will be kept at a minimum for years.

REPAIRING A LEAKY SINK STRAINER

1 Wear eye protection for this project. With the water turned off, beginning under the sink, remove the tailpiece (page 47). Next, loosen the locknut. Work with a screwdriver and hammer; tap lightly to loosen the lugs being careful not to damage the sink. Remove the locknut, the washer, and the rubber gasket. Pull the strainer body out of the sink.

For a retainer type strainer, simply unscrew the three screws and then disassemble as above.

2 Thoroughly inspect the washer and rubber gasket. If signs of wear are evident, replace them with duplicate parts. Thoroughly dry and clean the area around the drain opening and remove all old putty from the flange of the strainer. Apply a ⅛-inch bead of plumber's putty around the flange of the strainer. Place it in the sink opening and press down firmly to spread the putty and make a tight seal.

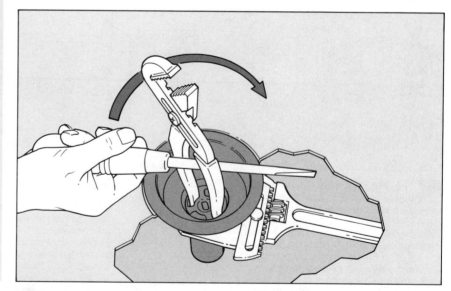

3 If the strainer is held in place by a locknut, it is best to work with a helper. From beneath the sink place the rubber gasket and washer around the strainer body. Have your helper hold onto the strainer body from above while you tighten the locknut from below with a spud wrench. Insert the handles of a pair of pliers into the slots of the strainer, wedge a screwdriver between the handles and have the helper hold it in place while you continue tightening. Screw on the coupling, reconnect the tailpiece, and check for leaks.

For a retainer type strainer, one person can finish the job. Have the gasket and washer firmly in place; put the retainer over the strainer body and tighten the three screws. Raise the coupling, screw it back on, and check for leaks.

Problems with Lavatory Pop-Ups

Old-fashioned sinks had rubber stoppers dangling from metal chains; they required a good shove to plug up the sink and a yank to remove. These have been replaced by pop-up stoppers (called simply *pop-ups*), mechanisms that raise and lower a stopper with a simple push-pull of a lift rod. Unfortunately, these convenient devices are not without problems of their own.

The most prevalent pop-up malfunction is a poor connection between the stopper and the sink. There are several possible causes for this. The rubber seal at the stopper might be worn or the flange or flange putty could be cracked. Often, the assembly needs to be adjusted. This procedure can be tricky because if you get the stopper too tight, then the sink won't drain quickly.

Another problem with pop-ups is leaks. If water is dripping near the pivot ball, it can be remedied by tightening the retaining nut that holds the ball in place or by replacing the gasket or washer (or both) inside the assembly. If water is dripping at the drain body, it is usually due to corrosion of one of the drain parts. In this case, replace the entire drain assembly (page 46).

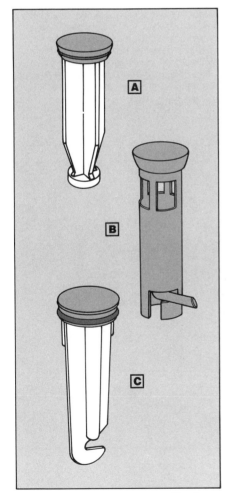

Types of Pop-Up Stoppers. The first step in adjusting a stopper is to remove it. This is done differently depending on the type. From top to bottom, remove by: (A) simply lifting out while the plug is opened; (B) slightly twisting to free it from the pivot rod; and (C) freeing the pivot rod (step 2, right) and pulling it out.

REPAIRING A FAULTY POP-UP STOPPER

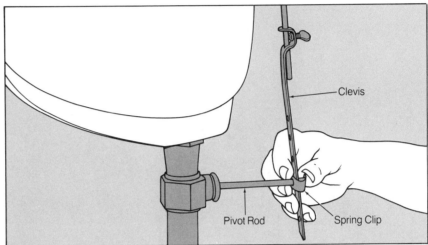

1 To remedy the problem of a plug that does not seal properly in the closed position, begin by adjusting the lift rod. Loosen the setscrew with a pair of tape-wrapped pliers. Press the pop-up down to seal the drain; this will cause the rod to rise. Pull the knob at the sink up as far as you can and then lower it only ¼ inch. In that position, tighten the setscrew. If the pop-up is now jammed or difficult to operate, then squeeze the spring clip tightly in your fingers and pull the pivot rod out.

Move the rod up to the next hole in the clevis and reassemble in the same manner, with the rod going through the clip and the hole. Test the pop-up stopper and if it is still not operating properly, start all over and adjust the lift rod in a different position.

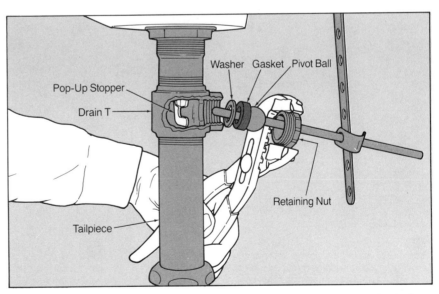

2 To free a pop-up stopper that is connected to a pivot rod, you must first unscrew the retaining nut; do this with a tape-wrapped adjustable wrench. Then squeeze the spring clip with your fingers and pull the pivot rod back out of the drain T.

To stop leaks from around the pivot ball, simply tighten the retaining nut. If you still have leaks, pull apart the assembly and check the plastic gasket and rubber washer; replace them if necessary.

REPLACING A DRAIN WITH A POP-UP

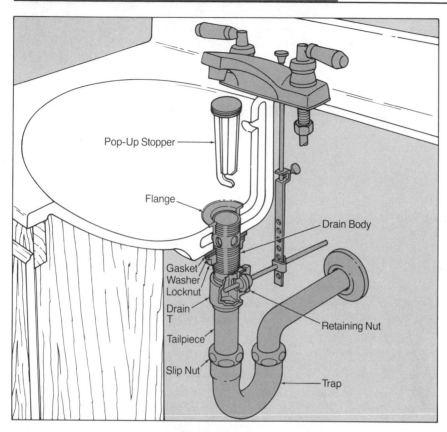

Pop-Up Stopper

Flange

Drain Body

Gasket
Washer
Locknut

Drain
T

Tailpiece

Retaining Nut

Slip Nut

Trap

Free the pivot rod from the drain (step 2, page 45). Loosen the slip nut that connects the tailpiece to the trap. Unscrew the tailpiece from the drain body and push it down into the trap. Remove the locknut-type strainer (step 1, page 44). Scrape away the putty at the mouth of the drain and wipe clean with a cloth. Discard the tailpiece but keep the slip nut that fastened it to the trap. If necessary, replace the slip-nut washer.

Install all parts using tape-wrapped wrenches. Begin by putting a slip nut and washer on the new tailpiece; push the piece down into the trap. Under the rim of the flange, apply a ⅛-inch bead of plumber's putty. Press the flange into place in the mouth of the drain. Coat the threads of the drain body and tailpiece with pipe-joint compound, screw the drain body into the flange. Next, screw the tailpiece into the drain body; tighten the tailpiece slip nut onto the trap. With an adjustable wrench, carefully tighten the locknut against the washer and gasket. Do not overtighten; the porcelain above it is liable to crack.

If you are also replacing the pop-up mechanism, feed the pivot rod into the drain T and tighten the retaining nut. Insert the lift rod downward through the faucet body and fasten its lower end to the clevis with the clevis screw. Insert the pivot rod through a clevis hole using the spring clip. Test the mechanism and if necessary, make adjustments (step 1, page 45).

INSTALLING A NEW SWIVEL P TRAP

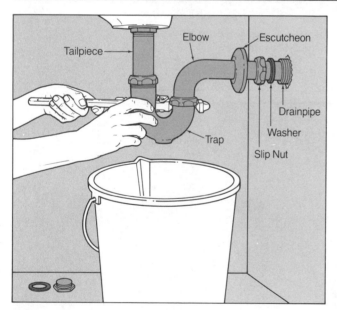

Tailpiece

Elbow

Escutcheon

Drainpipe

Washer

Slip Nut

Trap

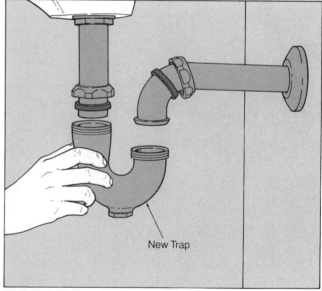

New Trap

1 With a bucket beneath the trap, remove the cleanout plug and gasket to empty standing water. With a tape-wrapped adjustable wrench, unscrew the two nuts that hold the trap to the tailpiece and the elbow. If the elbow needs to be replaced, pull the escutcheon from the wall and unscrew the slip nut that attaches the elbow to the drainpipe. If you are replacing the elbow, slide the escutcheon, a slip nut and a washer—in that order—onto the end of the elbow.

2 Put a thin coat of plumber's grease on the threads of the drainpipe. Tighten the slip nut and then push the escutcheon against the wall. Slide the slip nuts and washers for the new trap onto the tailpiece and elbow, as shown. Fit the trap into place and then tighten the slip nuts by hand. With a tape-wrapped wrench, continue tightening but be careful not to overtighten; this will strip the threads. Turn the water back on and check the connections for leaks.

Replacing Tailpieces and Traps

By the very nature of their design and function, traps are destined to wear out quickly. These bent pipes beneath sinks and lavatories serve the vital function of stopping sewage gases from traveling back into the house. But because water is left to stand inside these relatively thin pipes, corrosion is inevitable.

There are several types of traps. (For a thorough explanation of traps and venting, see page 76.) P traps are usually used in home plumbing, and there are some variations of them depending on how many pieces they are composed of and which pieces are movable.

If your sink is leaking at its trap, the cause could be stripped fittings or corrosion. If a sink continues to clog after repeated unclogging, the problem is usually mineral buildup. The solution for any of these problems is to install a new trap and possibly a new tailpiece. Tailpieces should be replaced if they are cracked or corroded.

There are a few criteria for choosing a new trap. First, of course, you should make sure that your fixture will comply with your local code. Next, make sure that the material is compatible with the pipes that it will connect to—namely, the tailpiece and the drainpipe. Finally, purchase the heaviest and finest material that you can afford. Chrome-plated traps are recommended; they are both good-looking and long-lasting.

Fixed traps screw directly onto the drainpipe making them awkward to work with. Swivel traps are easier to use. They can be turned in any direction on a drainpipe, allowing you to replace a tailpiece without removing it or to make a connection between a sink and drainpipe that are not perfectly aligned. Though fixed traps are not as versatile, you can replace a tailpiece without having to remove them. This is called 'rolling the trap' by plumbers. Swivel traps and fixed traps are both the same when it comes to installing a new drainpipe; in this instance, they both must be removed.

Tub and Shower Troubles

Tub faucets are like sink faucets in that they can also be of two kinds: compression and washerless. Use the instructions on pages 61-71 for making their repairs. However, tub faucets differ in a few minor ways. Because they are set into the wall, there is usually a problem gaining access for the stem repair. Also, the diverter valves on bathtub plumbing are unlike those on sinks.

INSTALLING A NEW TAILPIECE

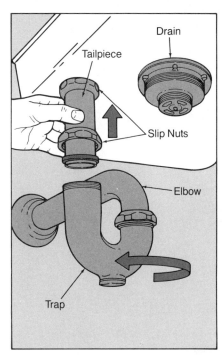

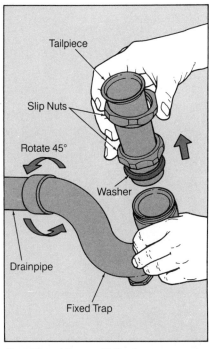

On a Swivel Trap. Wear eye protection for this project. Unscrew the slip nuts that hold the tailpiece to the trap and the sink drain. Push the tailpiece down into the trap. Loosen the slip nut that holds the trap to the elbow and swivel the trap away from the sink drain. Pull out the old tailpiece and insert the new one, coating the threads with plumber's grease. Swivel the trap back around, raise the new tailpiece into place, and tighten all slip nuts. Do not overtighten.

On a Fixed Trap. Using the same method as for a swivel trap, free the tailpiece and push it down into the fixed trap. With your hands or a tape-wrapped wrench, turn the entire trap assembly counterclockwise from the drainpipe. You only need to swivel it slightly—to an angle of about 45°. The old tailpiece can be removed and replaced. Reassemble, first the trap to the drainpipe, and then the tailpiece to the trap and drain. Tighten all slip nuts, but be careful not to overtighten.

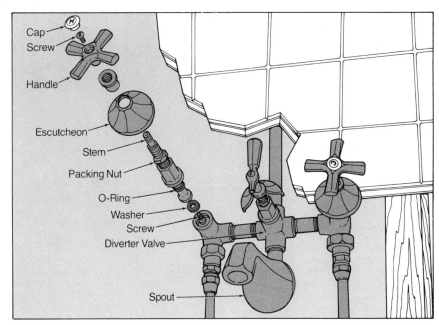

Anatomy of a Compression Tub Faucet. Tub faucets are like lavatory faucets and require the same repair procedures. Note that this type has a washer and packing in the form of an O-ring. Also, note the position of the diverter valve which directs water to the shower head or tub.

REMOVING A WALL-MOUNTED TUB FAUCET

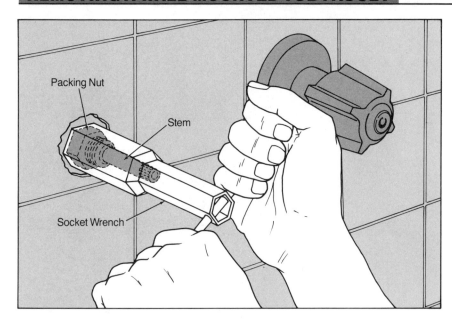

Packing Nut

Stem

Socket Wrench

In order to repair a compression tub faucet, you must first get at the packing nut which is sometimes set in plaster behind tile. Do this by chipping away a small section of the wall or tile. Chip away only enough to fit a deep socket wrench over the packing nut. Secure the wrench over the faucet stem and packing nut; turn to loosen.

CAUTION

Before working on tub faucets or diverters, turn off the water at the fixture valves or the main shutoff valve (page 9). Turn on the faucet and allow all water to run out.

REPAIRING TUB DIVERTERS

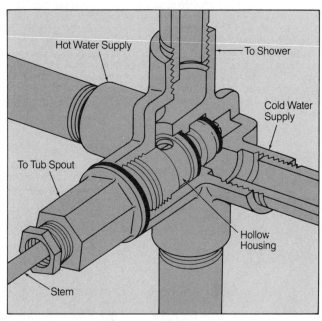

Hot Water Supply

To Shower

Cold Water Supply

To Tub Spout

Hollow Housing

Stem

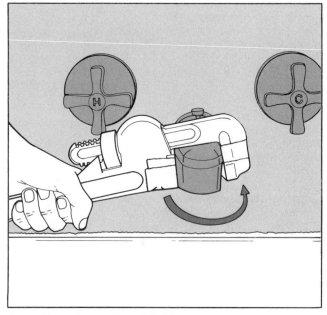

Repairing a Wall-Mounted Tub Diverter.   To remove the diverter valve, use the same method as was used for removing a wall-mounted tub faucet (above). The diverter valve works as follows: a clockwise turn closes the pipe to the tub spout and forces water through a hollow plastic housing and up into the shower head. A counterclockwise turn moves the stem back and opens the pipe to the tub spout. If the valve is leaky, disassemble it using the same steps as for disassembling a compression faucet (page 62). Replace any worn washers, O-rings, packing, or metal parts. If the flow of water is still not completely diverted to the shower head, the cause is probably a worn out hollow housing. Replace the diverter valve with a completely new assembly.

Replacing a Spout-Type Tub Diverter. This diverter is located in the tub spout; it controls the water flow when the knob is lifted or lowered. If it fails to work properly, the entire assembly must be replaced. To remove the spout, grip it with a tape-wrapped wrench and turn counterclockwise. An alternate method is to use a hammer handle inside the spout for leverage.

   If you can't find a spout of the same size you must use a nipple to extend the pipe behind the wall. Seal it with pipe-joint compound and screw the spout into the nipple by hand. If necessary, tighten by using a wrench or a hammer handle.

To remove either a wall-mounted faucet or a diverter valve, you first have to remove the bonnet nut which may or may not be recessed. If it isn't recessed, you can use a basin wrench or locking-grip pliers for the removal. If it is recessed, you'll have to use a deep socket wrench, available at plumbing supply stores.

There are two types of tub-spout diverters. One functions like a compression faucet and should be repaired in the same way, by examining and replacing any parts that are worn. The other type, which is incorporated into the tub spout, can only be repaired one way—by completely replacing the entire tub spout.

Shower head problems are numerous but usually very minor. Parts might simply be loose and need retightening, or they might be dirty and need scrubbing or soaking in vinegar. If you have a drip at the shower head, replace the washer or O-ring that fits around the swivel ball. If many of the parts are badly corroded and you want to make an improvement, consider replacing the entire unit with a hand-held shower (page 104).

Tub drains come in two varieties: a trip-lever type and a pop-up type. These mechanisms are hidden behind tubs and accordingly they are made of sturdy materials. Repair problems due to worn parts are very infrequent. The most common problem is the accumulation of hair on the trip-lever drain plunger or on the spring at the end of a pop-up lift linkage. Hair removal is accomplished by removing the lift linkage from the overflow tube.

Unlike sink pop-ups, tub mechanisms rarely need adjusting. When they do, use the same process to remove the lift linkage and then slightly raise its threaded rod. A tub pop-up stopper sometimes has a worn O-ring which is easily replaced. Pushing the stopper back into place is tricky but can be accomplished with a little patience.

Anatomy of a Trip-Lever Drain.
When you raise the lever of this mechanism the brass plunger is lowered to a ridge just below the junction of the tub drain and the overflow tube. Water is thus cut off from the tub drain.

However, since the plunger is hollow, any water that flows through the overflow tube will pass through the plunger and out the vertical drainpipe. If such a drain leaks, the cause could be a worn plunger. This can be remedied by simply lengthening the lift linkage. Another cause could be corroded cotter pins; replace them if necessary. After removal of the overflow plate, the assembly can be lifted out and worked on.

REPAIRING A FAULTY SHOWER HEAD

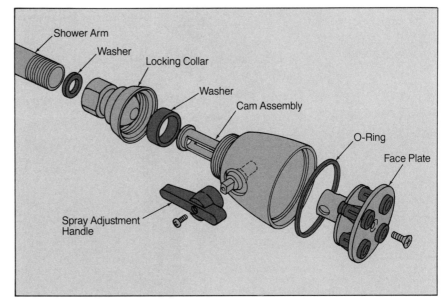

This type of shower head has a locking collar which means that you can remove it only by first unscrewing the connection at the shower arm. Before replacing a leaky shower head, first tighten all the connections and test. If that doesn't stop the leak, replace the washer between the swivel ball and the shower head. If the problem is sluggish water flow, there might be blockage in the face plate (or screen, depending on the model). To remedy this problem, clean all parts with a toothbrush, with a toothpick or, to remove mineral deposits, by soaking them in vinegar. Keep the parts in order for easy reassembly. Before reassembling, put a light coat of petroleum jelly on the swivel ball. Clean the threads of the shower arm stubout and apply pipe-joint compound before reinstalling the old or installing a new shower head. Tighten the shower head by hand only.

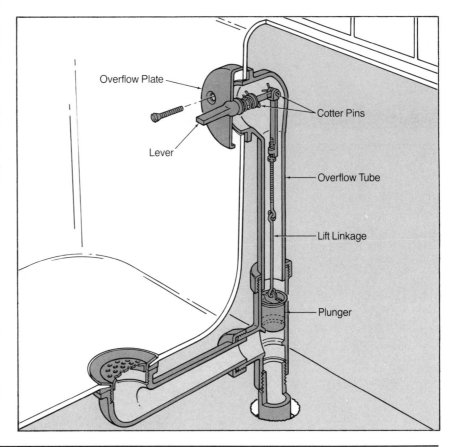

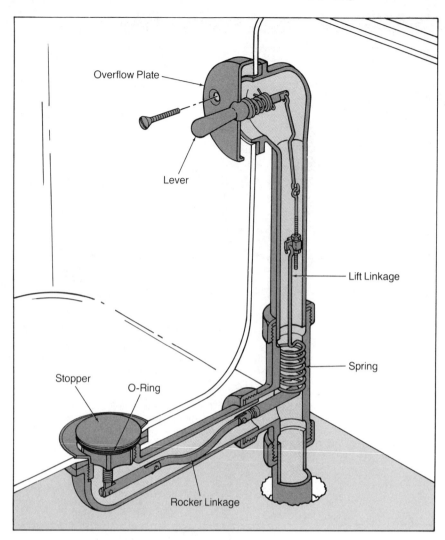

Anatomy of a Pop-Up Tub Drain.
Similar to the lift linkage of a trip-lever drain, the lower end of this linkage forms a stiff spring. The spring rests on the end of a horizontal linkage which leads to the stopper. This linkage is curved in the middle and is called a *rocker* linkage. When you push the lever up, the spring is pushed down and the stopper is raised. There is no strainer or screen at the tub outlet to prevent small objects from going down the drain. Instead, the design of the stopper, with wedges at its base, serves this purpose. People with sluggish drains sometimes try to solve their problem by completely removing the stopper; this is unwise because it can cause even more serious clogging. Instead, you should clear the tub drain with a snake (page 40).

ADJUSTING THE LIFT LINKAGE OF A TUB DRAIN

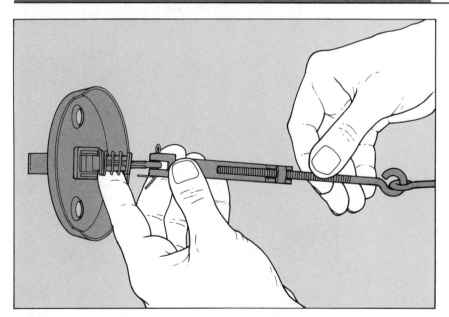

At the upper part of the lift linkage, there is a brass or plastic yoke inside of which is a threaded rod. The rod may or may not be held in place by a locknut. If it is not, simply hand-tighten the rod, as shown. If it is, use pliers to first loosen the locknut. Raise the rod slightly. Test it by running water in the tub and then noting how it drains. If necessary, make another slight adjustment. A linkage that has been lengthened too much will seal the drain properly but will not lift the plunger clear of the drain when the lever is in the open position.

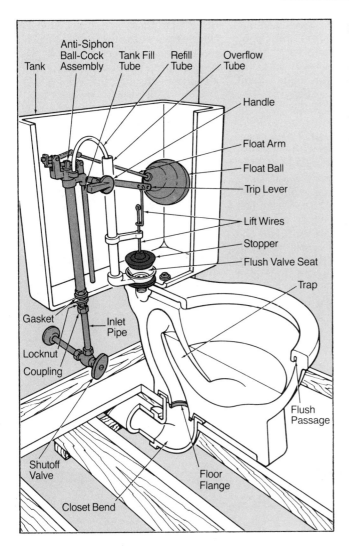

Tank

Anti-Siphon
Ball-Cock
Assembly

Tank Fill
Tube

Refill
Tube

Overflow
Tube

Handle

Float Arm

Float Ball

Trip Lever

Lift Wires

Stopper

Flush Valve Seat

Trap

Gasket

Inlet
Pipe

Locknut

Coupling

Flush
Passage

Shutoff
Valve

Floor
Flange

Closet Bend

Anatomy of a Toilet. Before attempting toilet repair, familiarize yourself with the parts and how they work. There are two basic assemblies: the ball-cock assembly, which regulates the filling of the tank and the flush valve assembly, which controls the water flow from the tank to the bowl.

Easy Toilet Repairs

Household tank toilets get a lot of use; in fact, almost half of the water used in a house passes through this mechanism. So it's no wonder that they have their little operation quirks. Homeowners have two reservations about toilet repairs: the water might look dirty and the machine looks complicated. The first misconception should be dismissed because tank water is as clean as anywhere else in your home, although the tank itself might look dirty due to sediments. As for the workings of the device, a brief study of its anatomy and how it works will allay your concerns and give you an understanding of just how simple it is. Study the drawing (shown at left) and note the toilet parts as you read the following description.

How a Toilet Works. The handle is pressed causing the trip lever to raise the chain or lift wires connected to the stopper. When the tank stopper goes up, water gushes out through the valve seat and into the bowl via the small flush passages lining the top of the bowl. The force of gravity pulls, or siphons, the water in the bowl out through the trap and through the drainpipe. After the tank has emptied, the stopper drops into the flush valve seat.

The float ball, now down, trips the ball-cock assembly and permits a new water supply to enter the tank through the tank fill tube. The water level rises, raising the float ball. When the ball reaches a certain height, it shuts off the flow of water. If the water fails to stop running, the tank will not overflow because water will go into the overflow tube and then into the toilet bowl.

Making Toilet Repairs. If your toilet bowl is clogged, see the emergency repair section on page 38 for how to handle the situation. For problems with the toilet mechanism, first remove the tank cover, flush the toilet several times and observe how it's working; often you can diagnose the problem merely by watching it in action. Then use the troubleshooting guide (on page 52) to make repairs, one by one, until you have eliminated the problem. If there are leaks from the tank or bowl or if the tank is 'sweating', refer to page 57 for the repair procedures.

REPAIRING A TUB POP-UP

A common cause of leaking pop-up drains is a worn O-ring. Open the drain and pull out the stopper along with the rocker linkage. Remove accumulated hair from these exposed parts. Remove the old O-ring and replace it with an exact duplicate. Push the stopper back into place, jiggling it from side to side to clear the bend in the pipe. The bottom of the curve of the rocker should face down.

Running Toilets. A toilet that continues to run long after you've flushed it can be quite a water-waster. Sometimes a slight jiggle of the handle will solve the problem but it's difficult to educate all members of the family, as well as guests, to perform this mini-chore. A much wiser choice is to make the repair immediately on discovering it, often a very simple process.

Sometimes the toilet continues to run because the water level is too high. An easy adjustment can be made by bending the float arm to lower the float. Do this carefully, with both hands, to avoid damaging any other parts of the assembly. Another possibility is that the float ball is obstructed. Check to make sure that the ball is not rubbing the back or side of the tank.

Even if you don't hear the sound of trickling water, your toilet might be prone to 'silent' leaks. These are caused by corroded flush valve assemblies and they can sometimes be remedied by simply cleaning the valve seat.

After turning the water off at the fixture valve or at the main shutoff valve, check the valve seat for corrosion. Seats, usually made of copper, brass, or plastic, are susceptible to mineral build-up. If the surface does not look clean, gently scour the seat with fine steel wool.

Leaks are caused by one or more parts of the ball-cock assembly. These come in many styles. The most common is the *plunger* type, which consumes about 5 gallons of water per flush. The *float-cup* ball-cock is a simpler-designed unit that contains no float arm or ball; 'silent leaks' are virtually eliminated with this type.

If you need to replace the stopper—the bottom part of the assembly that drops into the valve seat—you'll dis-cover that there are many kinds to choose from at the hardware store. It's recommended that you select a flapper-type stopper, a model that's less prone to misalignment.

Ball-cock assemblies have small washers that can be replaced (page 55) but if many parts of this assembly are worn you should consider making a total replacement. If you wish to update the system, it is recommended that you choose the most modern replacement available. Besides having an anti-siphon feature, a requirement of most codes, the new models are easier to install.

> **PLUMBER'S TIP:** Installing a new ball-cock is one of the most effective repairs that the home plumber can make.

TROUBLESHOOTING TOILET PROBLEMS

TROUBLE	POSSIBLE CAUSE	SOLUTION
Water continues to run long after flushing.	Float arm isn't rising to correct height.	Adjust the float arm by bending.
	Stopper isn't seating properly into flush valve.	Adjust the chain, rods, or wires. If necessary, replace the stopper.
	Flush valve seat is corroded.	Clean the valve seat or replace it.
	Float ball is filled with water.	Replace float ball.
	Overflow tube is cracked.	Replace overflow tube or install new flush valve assembly.
	Defective ball-cock valve.	Replace faulty washers, oil lever, or install new ball-cock assembly.
Insufficient flush.	Faulty linkage between trip lever and handle.	Tighten screw on handle linkage or replace handle.
	Tank stopper closes before tank is emptied.	Adjust stopper guide chain or rod.
	Flush passages are clogged.	Clean obstructions from passages by poking with wire.
	Leak between bowl and tank.	Tighten the locknuts under tank or replace gasket.
'Whistling' flush.	Ball-cock is defective.	Replace faulty washers, oil lever, or install new ball-cock assembly.
Leaking tank.	Loose fittings or worn washers.	Tighten hold-down bolts or replace washers. Replace flush-valve washers. For a wall-mounted tank, install new packing around slip nuts.
Sweating tank.	Condensation, normal when warm air meets cold water.	Install a tank liner or a temperature valve.

If a toilet's running is caused by a cracked overflow tube, you either need to replace the tube itself or install an entire new flush-valve assembly. This process (page 56) involves working under the tank.

Inadequate Flush. When the toilet fails to operate to its potential, it is either because the mechanism, usually at the handle, is not functioning properly or because there is a leak between the tank and the bowl. Follow the repair steps given in the chart, one by one, until you have located and solved the problem.

'Whistling' Toilets. When a toilet makes a high-pitched whistling or whining sound, the culprit is a ball-cock leak. The plunger of this assembly contains small washers which, if cracked, can make an inordinate amount of noise. Either replace the washers or, if necessary, install a brand-new ball-cock.

Toilets that Leak or Sweat. Too much water running inside of a toilet is one thing, but when water is running off of the toilet it can do damage to tiles or wood, or cause mildew to form. Diagnosis may be difficult between sweating or leaking but if the problem is around the hold-down bolts, there is an easy way to figure it out. Pour laundry bluing inside the tank and then hold a piece of tissue paper over the bolts. If blue water exposes a leak, tightening the bolts will usually cure the problem.

TYPES OF BALL-COCK ASSEMBLIES

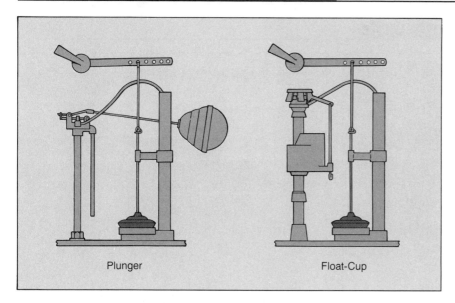

Plunger

Float-Cup

Ball-cocks allow water to pass into the tank when the toilet is flushed and shut off the water flow afterwards. Shown here are two popular types of ball-cock assemblies. The plunger type has a float arm which applies pressure on a valve and plunger to seal off the incoming water. The simpler, easy-to-install, float-cup type has no float arm or ball. The plastic cup or fill valve works to control the water flow.

IMPORTANT

There are many kinds of ball-cocks on the market. Whatever type you purchase, make sure that it is an *anti-siphon* model. Almost all codes require this type.

CORRECTING A FLUSH PROBLEM

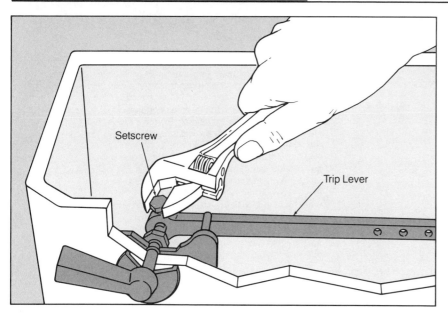

Setscrew

Trip Lever

One cause of incomplete or erratic flushes is a faulty handle and trip lever connection. With an adjustable wrench, tighten the setscrew on the handle linkage. Check for continuing problems. If they persist, adjust the stopper guide rod or chain, clean the flush passages under the rim of the bowl with a piece of wire, and also check for leaks between the tank and bowl. If these areas are satisfactory and you still suspect that the problem is in the handle connection, replace the handle.

Leaks around the flush valve, the large opening between the tank and the bowl, are usually due to worn washers. Their replacement necessitates removal of the tank, a fairly heavy fixture that you should handle carefully. Older wall-mounted toilets have L-shaped pipes that connect the tank to the bowl. Leaks at the connecting slip nuts can be stopped by loosening them, inserting packing and retightening. Leaks at the base of the toilet are likely caused by a worn wax gasket. Remove the bowl and replace the wax gasket using instructions for replacing a toilet, pages 98-100.

Sweating toilets are caused by condensation; when moist warm air meets with the cold tank surface, water is formed. If the water temperature is 50° or higher, you can eliminate the problem by installing a tank liner. These are available in kits at plumbing supply stores, or you can easily make your own from foam rubber purchased at a department store. If the water temperature is colder than 50°, the only remedy is to install a *tempering valve* (not shown), a relatively expensive fixture that mixes hot and cold water.

REPAIRS FOR 'RUNNING' TOILETS

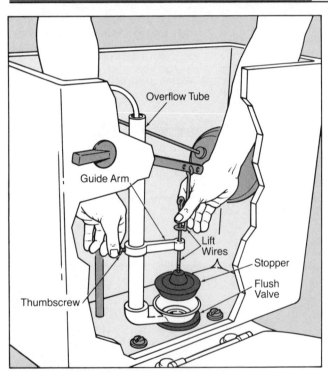

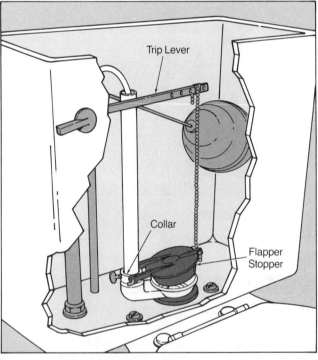

Adjusting the Tank Stopper. A common cause of running toilets is an imperfectly seated tank stopper. Begin this adjustment by shutting off the water to the tank. Observe the stopper as you flush the toilet. If it fails to drop directly onto the flush valve, loosen the thumbscrew that holds the guide arm to the overflow pipe. Adjust the lower lift wire and the guide arm so the stopper is centered over the flush valve. If necessary, straighten both the lower and the upper wires.

CAUTION

Before working on the toilet stopper or the valve seat, turn off the water at the fixture valve or at the main shutoff valve (page 9).

Installing a Flapper Stopper. If your tank stopper needs replacing, install a flapper-type stopper to eliminate future problems of alignment with the lift wires or guide arm. First, flush to drain the tank. Remove the old guide arm and lift wires. Put the collar on the overflow pipe and slide it to the bottom. Position the stopper over the outlet valve and then tighten the thumbscrew on the collar. Hook the chain to one of the holes in the trip lever; leave about ½ inch slack. Turn the water back on, flush the toilet and watch to see if the tank drains completely. If it does not, tighten the chain by moving it to a hole further toward the handle or by lessening the slack.

ELIMINATING A BALL-COCK LEAK

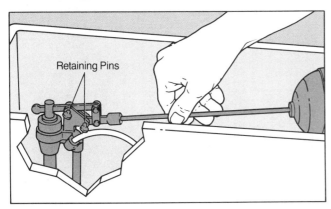

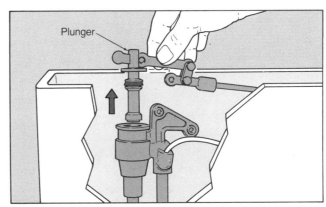

1 Lift the float arm as high as it will go and observe the action of the ball-cock. If water continues to run in, the problem could be defective washers on the plunger. Turn off the water for the repair. Pull out the two retaining pins that hold the assembly.

2 Pull the plunger up and out of the ball-cock assembly.

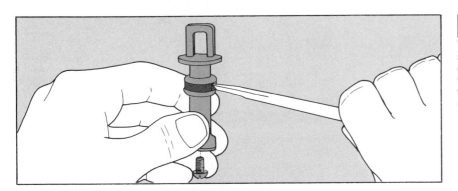

3 Examine the washers; these tiny parts can cause loud tank noises or leaks. There will be a seat washer and one or more split washers. Remove them all with a screwdriver and then scrape any mineral residue from the plunger being careful not to scratch the metal. Replace the washers with exact duplicates.

INSTALLING A NEW BALL-COCK

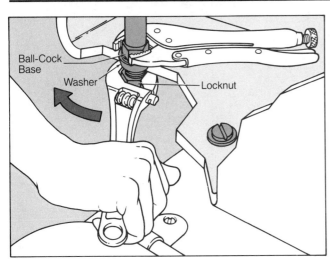

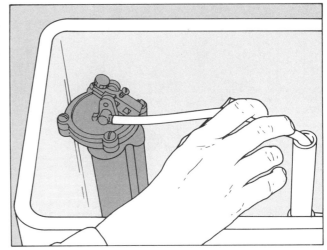

1 If you have replaced the plunger washers and still have leaks, the next step is to replace the ball-cock. Choose a newer, quality model for this replacement. Check to make sure that it is an anti-siphoning type ball-cock.

Begin the disassembly by turning off the water and flushing the tank twice. Then sponge out any remaining water. Remove the float mechanism. With an adjustable wrench grip the coupling that holds the water-inlet pipe on the underside of the tank. Inside the tank, use locking grip pliers or an adjustable wrench to hold the base of the ball-cock while detaching the locknut. If it is resistant, soak in with penetrating oil for 15 to 20 minutes and then try again. When the locknut is removed, lift the ball-cock out of the tank. Carefully check the coupling and gasket and replace them if worn.

2 At the bottom of the new assembly, slip on a washer and locknut, in that order. Tighten the nut to secure the assembly. Once installed, tighten the coupling and gasket on the inlet pipe under the tank. Position the bowl-refill tube inside the overflow tube. Assemble the float arm and the ball and attach them to the ball-cock. Make sure that the locknut and the coupling are tightened securely. Turn the water on to fill the tank and adjust the float arm.

REPLACING A FLUSH-VALVE ASSEMBLY

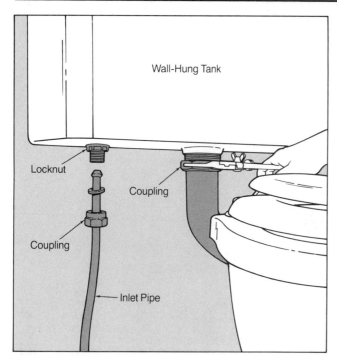

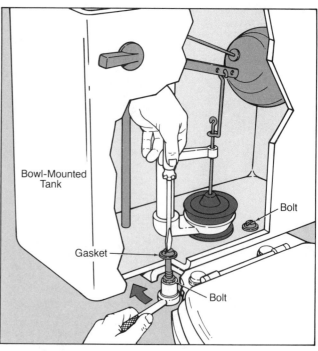

1 Flush the toilet to empty the tank. Remove the old guide rod, stopper, lift wires, or chain. If your toilet is the wall-hung type, loosen the coupling on the bend of pipe that connects the tank to the bowl. If necessary, loosen the coupling at the other end of the pipe. Remove the pipe. Unscrew the locknut that attaches the inlet pipe.

If your toilet has a bowl-mounted tank, remove the bolts and the gaskets, as shown.

CAUTION

Before working on the flush valve assembly, turn off the water at the fixture valve or at the main shutoff valve (page 9).

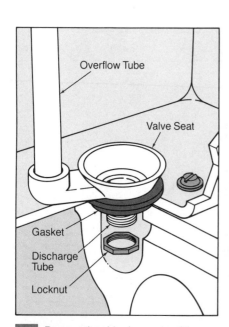

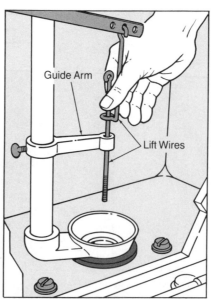

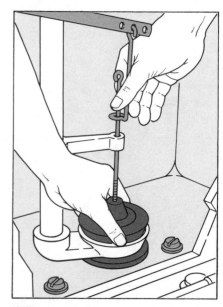

2 Remove the old valve seat and the gasket. At the bottom of the tank, insert the new discharge tube. Position the new overflow tube and tighten the locknut beneath the tank to hold it in place.

3 Position the guide arm on the overflow tube directly over the valve seat and tighten the thumbscrew to secure it. Install the lift wires through the guide arm and then through the trip lever.

4 Screw the tank stopper onto the lower lift wire and carefully align it directly above the valve seat. (If you're installing a flapper-type stopper, use the instructions on page 54.)

STOPPING LEAKS

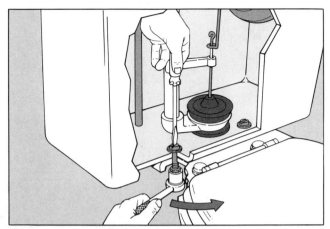

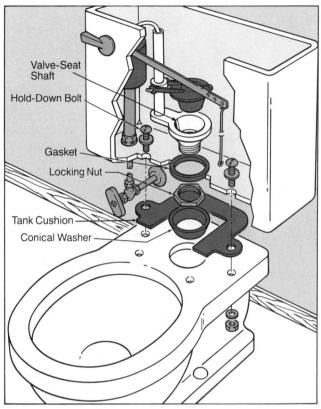

At the Hold-Down Bolts. After draining the tank, have a helper hold the head of the bolts inside the tank with a screwdriver. Tighten the nuts below the tank using an adjustable wrench or a socket wrench. Do not overtighten as you could crack the finish of the tank. If this doesn't stop the leak, remove the bolts and replace the washers.

Valve-Seat Shaft
Hold-Down Bolt
Gasket
Locking Nut
Tank Cushion
Conical Washer

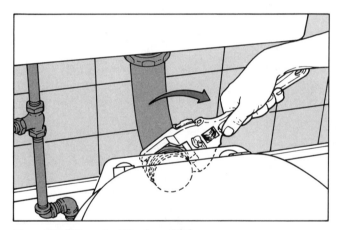

From Wall-Mounted Tanks. Leaks occur at the fittings of wall-mounted tanks due to shifts in walls or floors. Use a large wrench or spud wrench to unscrew, counter-clockwise, the couplings at the tank and the bowl. Wrap self-forming packing around the exposed threads and then retighten.

At the Flush Valve. If leaks occur around the flush valve, you must remove the tank. Drain it completely and unscrew the hold-down bolts. Also, disconnect the supply pipe to the ball-cock. Lift the tank upward to remove. Unscrew the large locking nut that connects to the valve-seat shaft. Pull the shaft up into the tank. Replace its gasket and its large conical washer. Remount the tank and check for leaks.

CAUTION

Before working on the flush valve assembly, turn off the water at the fixture valve or at the main shutoff valve (page 9).

SWEATPROOFING A TOILET TANK

To make your own tank liner, measure for dimensions inside the tank and cut pieces of ½-inch-thick foam rubber to fit. Check carefully to make sure that the lining won't interfere with moving parts and make cutouts if this is the case. Apply a liberal coating of rubber cement or silicone glue to a dry tank. Press the foam into place and let stand for 24 hours before using.

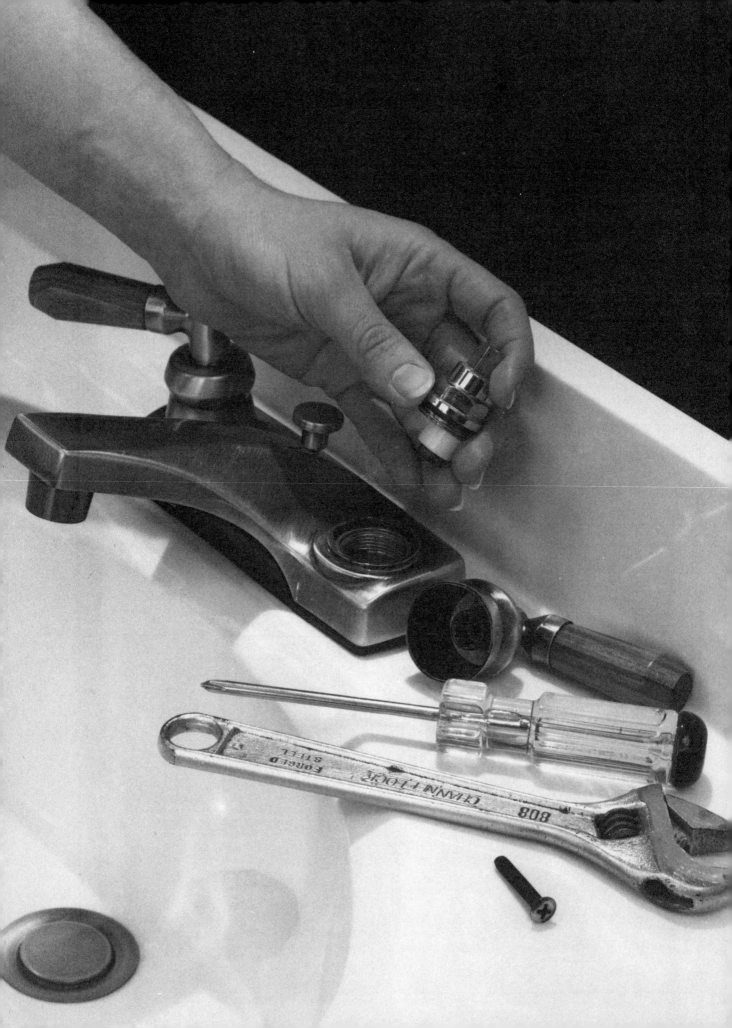

Faucets and Valves

Your life might be so hectic that the only time you hear the drip-drip of your leaky faucet is late at night when your other cares have melted away. The sound can be quite harassing but apart from that you should also consider the pennies that are rolling down the drain. One drop of water every two seconds can add up to a gallon a day, certainly enough to eventually swell up your water bill. Moreover, constant leaks will stain and erode fixtures.

The topic of faucets, which is potentially mind-boggling, is presented here in a logical, easy-to-understand manner. Though there are dozens of styles, only the most popular ones are shown. If what you see when you disassemble your faucet doesn't look exactly like what is depicted in this chapter, don't panic. Parts vary in size and shape but they generally remain similar in function.

One trick is to always lay your faucet parts out in the order that you removed them, or, if you want to take several parts to the hardware store with you, number them for reference in reassembly. As a footnote to faucet repair, the home do-it-yourselfer is in luck, since nowadays many products come with how-to diagrams and instructions. When installing a new faucet, be sure to store all such paperwork in a home-repair file.

Calling a plumber to repair a faucet seems to be quite extravagant, but it's also a home chore that often ends up on the 'put-off-til-tomorrow' list. Truthfully, the only crucial part is to make sure that your water is shut off. Aside from this, with patience and the minor inconvenience of having no water for a while, practically anyone can repair their own faucets.

Faucets and Valves— Victims of Use

Perhaps the single most common plumbing repair is on faucets, and for good reason. When you think about how many times a day you turn the water on and off, it only makes sense that these little devices are pushed to the limit. Moreover, supply water lines are continually under pressure so a faucet or valve is working even when it's not working!

The subject of faucets is a little complex, since there are so many different kinds. Simpler in construction than faucets are valves, such as those used at shutoffs for fixtures and appliances. There are three basic kinds of valves, depending on how they're used. They are usually made of cast bronze although plastic and brass valves are also made. (Brass valves are typically used in gas systems.) Valves are very basic mechanisms, consisting of a handle that drives down a stem into its base.

Even though valves aren't used quite as often as faucets, they do wear out—sometimes as a result of being used incorrectly. So even if you have valves that aren't in need of repair now, it is wise to identify them and understand how they work. When a valve wears out it will leak in its stem.

Types of Faucets— A Simplification

In order to repair a faucet, you must first identify it. This won't be quite as tricky as it first seems if you use the pictures, not the words, and understand a few basic principles. Part of the problem is that of vocabulary. Some types of faucets have two or three different descriptive names, all of them correct.

However, there are really only two basic kinds of faucets.

The first basic kind is the older-style *compression* faucet. Additional names for this are *washer* or *stem* faucet. Its chief characteristic is the washer that fits snugly into the valve seat and stops the flow of water. Compression faucets are easily identified on older sinks because they come in pairs, one for hot and one for cold water. A more recent variation of this has two separate faucets but a shared spout which allows hot and cold water to mix.

The second basic kind of faucet, and one generally requiring less repair, is the modern *washerless* or *single-lever* faucet. It consists of a single lever or knob that controls the flow and mix of hot and cold water by aligning interior openings with water inlets. When first introduced, this faucet was used exclusively at kitchen sinks because it so easily provided the desired water temperature; now it is a standard fixture for lavatories.

Washerless faucets come in four different types, depending on the inner mechanism: *ball, disc, cartridge,* and *tipping valve.* These might be difficult to distinguish from each other so you should always go by what is inside the faucet and not the exterior appearance. The most common is the cartridge type.

Once you have identified your faucet, you should read through all the repair steps so that you'll understand what the project entails. Also, be sure to read and follow the *Non-Emergency Repair Guide* on page 44 for some additional helpful tips.

HOW VALVES WORK

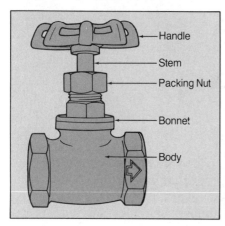

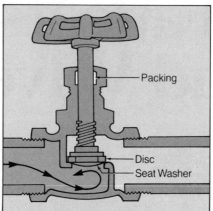

How a Globe Valve Works. Globe valves reduce water pressure because their interiors consist of two half partitions that slow the water down. Similar to a compression faucet (right), a globe valve has a disc that is forced by a stem into the valve seat. Globe valves can take frequent opening and closing under high pressure. The easiest kind of valve to repair, they usually have defective washers or seats that need to be replaced.

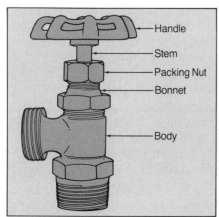

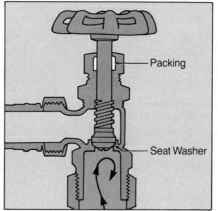

How an Angle Valve Works. Similar to the globe valve, the angle valve is designed with the water inlet and outlet at a 90° angle to each other. The angle valve has no half partitions to slow the flow of water. These are often used in place of an elbow fitting when pipe turns a corner.

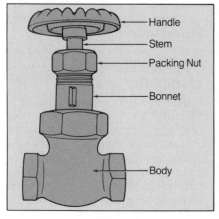

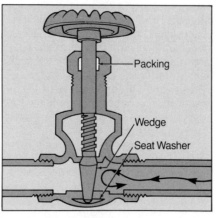

How a Gate Valve Works. In residential systems, gate valves are used to completely open or shut down supply pipes; they are not meant to be partially opened. When the handle is turned to close the valve, the metal wedge is pushed into the seat washer. If the handle is turned to completely open the valve, the metal wedge is out of the water's path. Trouble arises when the handle is only partially turned; this allows water to eventually erode the metal wedge, resulting in a leaky valve. With proper use, gate valves last for years.

Repairing a Compression Faucet. Many things can go wrong with a compression faucet; they are prone to leaks in several places. The most difficult repair task is often disassembly. Stems, the threaded rods of the faucets, are constructed and installed differently and you should follow a logical progression of steps to remove them. Once removed, you can begin your identification and inspection of parts.

Leaks most often occur at the spout and the culprit in this case is a worn seat washer or its mating metal seat.

When the spout is common to two faucets, then you first have to determine which one is causing the drip. Do this by turning off the supply valves under the sink until the drip stops.

There are several types of washers or washer-assemblies and they are all installed a little differently. It might be difficult to find the exact washer that you need and you might end up having to purchase an assortment package. If your leak continues after you have replaced the washer, then the problem is

in the valve seat which has developed a rough surface or burrs on its surface. It must either be replaced or, if it is not removable, renewed. This grinding process is done with an inexpensive tool, a valve seat dresser (page 12).

If leaks occur at the faucet handle when the water is turned on, then the problem is either a loose packing nut or the packing itself. There are three basic types: O-ring, washer, or twine; replacement of these parts is relatively easy.

REPAIRING A STEM LEAK IN A VALVE

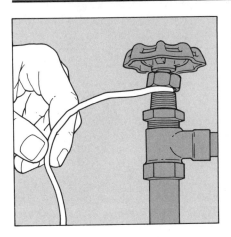

1 Begin by draining the system; turn off the water at the main shutoff valve and then run water from the nearest faucet to your work site. Remove the valve's packing nut, located just below the handle, with an adjustable-end wrench. Look closely at the packing; if it's compressed then you must remove all of it and replace it with new packing. Wrap the stem clockwise with new graphite-impregnated twine. Reassemble and then test for a leak.

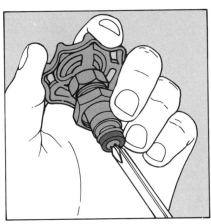

2 If replacing the packing fails to stop the leak, then you need to check further. Unscrew the valve stem and the bonnet from the body, pull out the assembly and check the seat washer at the bottom of the stem. If it looks worn, unscrew the setscrew and remove the washer. Replace it with a new one exactly like it.

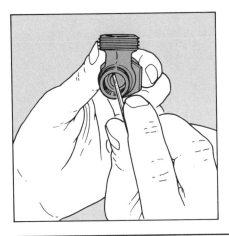

3 If the valve still leaks, check for an obstruction in the valve body. Remove the stem as described above, and insert a toothpick for probing and removing foreign matter. If necessary, unscrew the valve from its connecting pipe and clean the body with soapy water and a stiff brush.

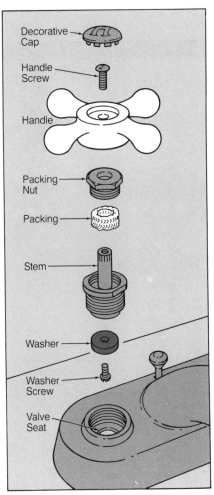

Anatomy of a Compression Faucet. A compression faucet works in this way: a screw pushes and compresses a washer against a valve seat. When the faucet is turned off, the stem is screwed all the way down and the washer fits snugly into the valve seat, stopping the water flow.

CAUTION

Before working on faucets, turn off the water at the fixture valves or the main shutoff valve (page 9). Turn on the faucet and allow all water to run out.

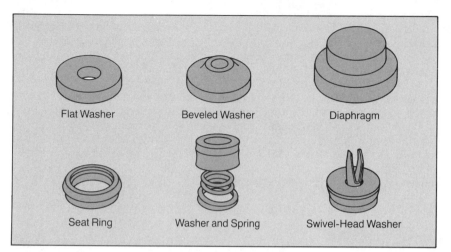

Flat Washer

Beveled Washer

Diaphragm

Seat Ring

Washer and Spring

Swivel-Head Washer

Type of Washers in Compression Faucets. Compression faucets are also called *washer faucets* because they have washers in their assemblies. The most common types of washers are flat or beveled with a hole in the middle for the washer screw. One type is actually a washer-like diaphragm that covers the bottom of the stem. The seat ring fits around the stem and acts like a washer. Washer and spring assemblies are found in cartridge-type compression faucets. If the washer screw breaks, you can install a swivel-head washer; it has two prongs that snap into the bottom of the stem.

TAKING APART A COMPRESSION FAUCET

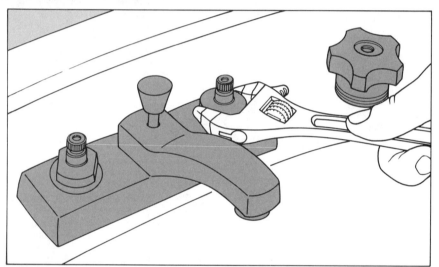

1 First, remove the handle screw. Usually this is hidden by a decorative button or cap. Pry off this cap with a sharp tool such as a utility knife, being careful not to mar surfaces. With a screwdriver remove the handle screw and then pull the handle straight up.

2 Next, using a tape-wrapped adjustable-end wrench, remove the packing nut which, in this faucet, is a locknut. Turn the wrench counterclockwise. The faucet stem might come out with the locknut. If it does, wrap it with tape, clamp it in a vise, and twist off the locknut with a wrench.

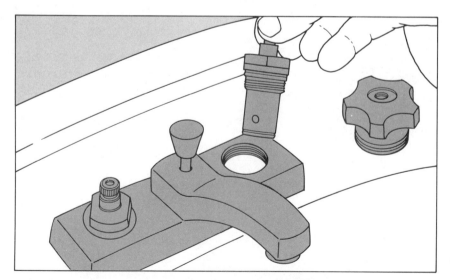

3 If step 2 does not provide you with the faucet stem, then replace the handle and turn it in the direction that would normally turn on the water; continue to turn until the stem comes out. If the faucet has a cartridge-type stem (page 64, bottom), pull it out by hand or wrap the stem with a rag and pull it out with pliers.

CAUTION

Before working on faucets, turn off the water at the fixture valves or the main shutoff valve (page 9). Turn on the faucet and allow all water to run out.

FIXING A HANDLE LEAK IN A COMPRESSION FAUCET

First, try to stop the leak by tightening the packing nut; do not force it too tight. If the leak persists, you need to replace the packing by one of these three methods, depending on the type of faucet.

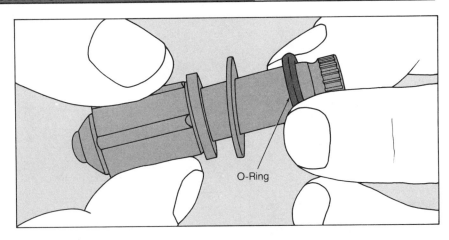

Replacing the O-Ring. First, remove the stem as instructed in steps 2 and 3, left. Then remove the old O-ring by pinching it tight with your fingers and pulling it off. Replace it with an O-ring identical in size and shape.

O-Ring

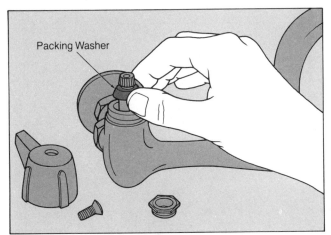

Packing Washer

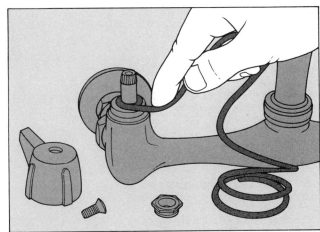

Replacing the Packing Washer.
First, remove the stem as instructed in steps 2 and 3, left. Remove the old packing washer and replace it with a washer identical in size and shape. If you cannot find an identical washer, you can use twine packing. Push it onto the stem as far as it will go. Before replacing the packing nut, lubricate the threads of the nut and stem with petroleum jelly. Reassemble.

Replacing the Twine Packing. This self-forming packing is available in the form of graphite-impregnated twine and is available at plumbing stores. One strand at a time, wrap the twine in layers around the faucet stem using about half again more than is needed to fill the packing nut. The packing will compress solidly once the nut is screwed down over it. Before replacing the packing nut, lubricate the threads of the nut and stem with petroleum jelly. Reassemble.

FIXING LEAKY SPOUTS OF COMPRESSION FAUCETS

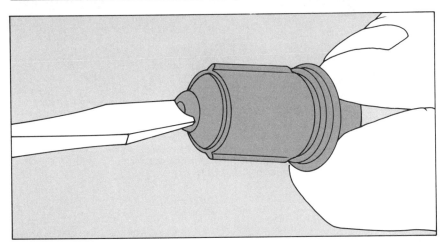

Replacing a Flat or Beveled Washer.
Remove the handle and stem using instructions at left. At the bottom of the stem is a seat screw that goes through the middle of the washer. If the washer is cracked, marred, or grooved, it should be replaced. Carefully remove the screw and replace the washer with one exactly like it. If the washer is beveled, install it on the stem with the beveled edge facing the screw head.

CAUTION

Before working on faucets, turn off the water at the fixture valves or the main shutoff valve (page 9). Turn on the faucet and allow all water to run out.

FIXING LEAKY SPOUTS OF COMPRESSION FAUCETS/CONT'D

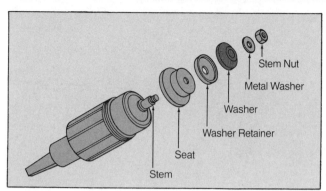

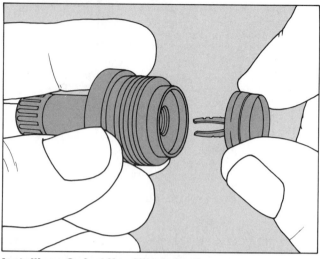

Replacing a Beveled Washer on a Reverse-Pressure Faucet. This compression faucet is constructed differently—the faucet seat is built into the stem instead of the faucet body. When the water is turned off, the washer moves upward to close against the seat—opposite of the direction a washer moves in a standard compression faucet. To replace the washer, unscrew the stem nut; remove it, the metal washer, rubber washer, and washer retainer. Take off the metal seat and examine it; replace it if its surface is worn. Put the new washer, beveled side up, into the retainer cup.

Installing a Swivel-Head Washer. If a washer screw is broken, instead of using a regular washer with a hole in it use a swivel-head washer of the same size. If it does not snap securely into the screw hole of the stem, place the stem in a vise and drill a longer or wider hole into the bottom. Snap the prongs into the hole.

Replacing a Diaphragm. Remove the handle and stem using instructions on page 62. Sometimes the old diaphragm sticks inside the faucet body; if so, pry it out with the tip of a screwdriver. Inspect the faucet body, using a flashlight; if all debris is not removed the new diaphragm will not seat properly. Fit the new diaphragm onto the bottom of the stem, making sure that it fits snugly.

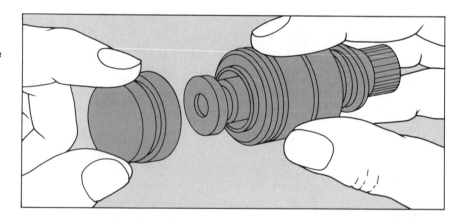

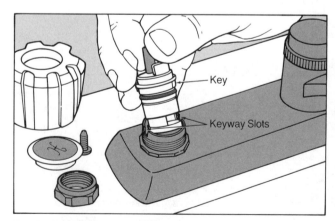

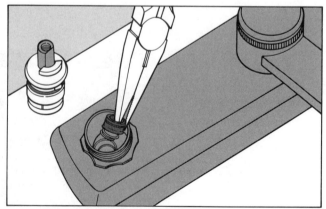

Replacing a Washer and Spring Combination. In this cartridge-type compression faucet, the cartridge presses against a separate washer and spring in the faucet body. Lift the cartridge out of the faucet noting the position of the top in relation to the keys on its side and the keyway slots on the faucet body.

Pull out the old washer and spring with needlenose pliers. Push in the new washer and spring with your index finger.

Replacing a Seat Ring. Hold the small rectangular part of the stem with needlenose pliers. Unscrew the threaded centerpiece which holds the seat ring. Remove the sleeve and replace the seat ring. The lettering on the new ring should face the threaded part of the stem.

CAUTION

Before working on faucets, turn off the water at the fixture valves or the main shutoff valve (page 9). Turn on the faucet and allow all water to run out.

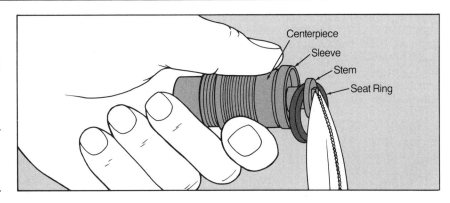

Centerpiece
Sleeve
Stem
Seat Ring

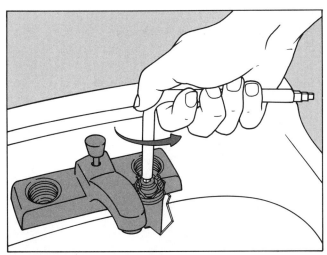

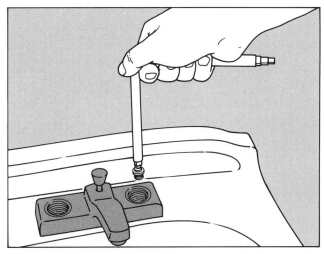

Renewing a Valve Seat. When a faucet continues to drip at the spout after you have replaced its washer, the problem could be a faulty seat. Inspect it with a flashlight; look for pits or scratches. Then run your fingertip along the surface of the seat, feeling for unevenness.

Most faucet seats are replaceable and installing a new one is much easier than repairing an old one. With a seat wrench, turn the seat counterclockwise; then lift it out. Replace it with an exact duplicate.

Put pipe-joint compound on the outside of the new seat, push it firmly onto the seat wrench and screw it into the faucet body.

If your faucet seat is not removable, you need to grind the surface smooth with a valve-seat dresser. Use the largest dresser that will fit the faucet body. Insert it until the cutter sits on the valve seat. Turn the handle several times, pressing lightly. Remove metal shavings with a damp cloth.

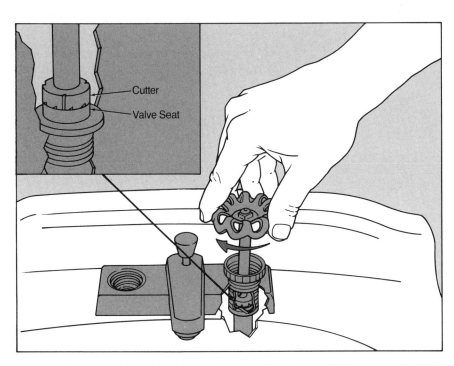

Cutter
Valve Seat

Repairing a Ball Faucet. If the spout of a ball faucet is dripping, the problem is usually the inlet seals or springs; they are worn and should be replaced. If the handle leaks, then the problem can be corrected by tightening the adjusting ring or replacing the cam seal above the ball. When the problem is a leak under the spout, you need to replace the O-rings or the ball itself.

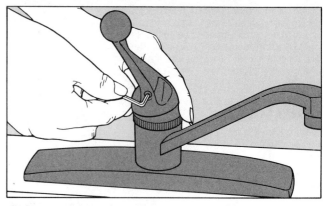

REPAIRING A BALL FAUCET

1 Begin disassembling the faucet by removing the setscrew with an Allen wrench.

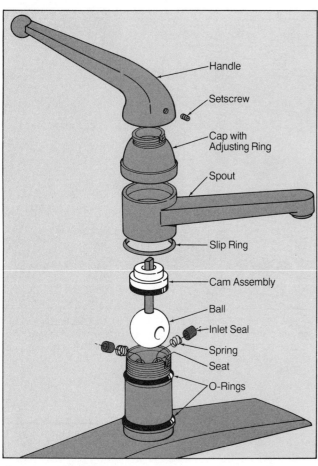

Handle

Setscrew

Cap with Adjusting Ring

Spout

Slip Ring

Cam Assembly

Ball

Inlet Seal

Spring

Seat

O-Rings

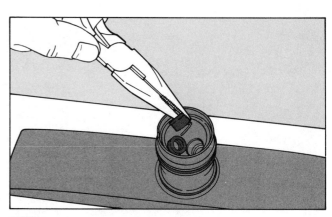

4 Remove the spout sleeve to expose the faucet body. With needlenose pliers, remove the two seals and springs. Remove any buildup in the inlet holes with a penknife or stiff brush. Check to see if new O-rings are needed. If so, replace them and apply a thin coat of petroleum jelly to the new O-rings. If necessary replace the seals and springs, pushing them into place with your fingertip.

Anatomy of a Ball Faucet. Ball faucets have a slotted metal ball atop two spring-loaded rubber inlet seats. When the openings in the rotating ball align with hot and cold water inlets, water begins to flow.

Four Kinds of Washerless Faucets. These are the four kinds of washerless faucets: ball, disc, cartridge, and tipping valve. Also called *single-lever* faucets, they all have a single lever or knob that controls the mix and flow of hot and cold water. Though all four look very different, do not determine what type you have based on the exterior alone; once you disassemble the faucet you can determine what kind of mechanism it has.

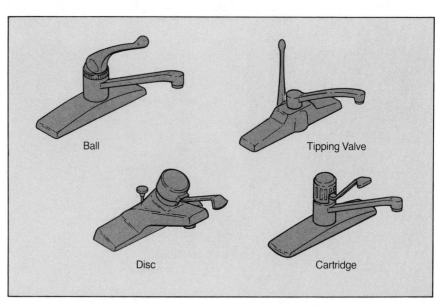

Ball

Tipping Valve

Disc

Cartridge

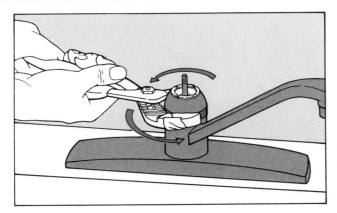

2 With tape-wrapped rib-joint pliers, unscrew the cap.

3 Lift out the stem and the ball; the plastic and rubber cam assembly will come with it. Underneath it will be two inlet seals on springs.

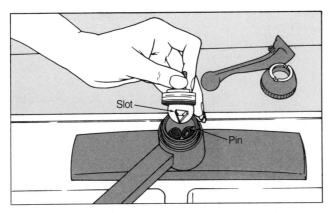

Slot

Pin

5 Reinstall the spout, then insert the ball. If the ball is rough or corroded, it should be replaced. When inserting a new ball, make sure that you line it up correctly. The ball has a slot in it that should be put over a pin that lies in the faucet body.

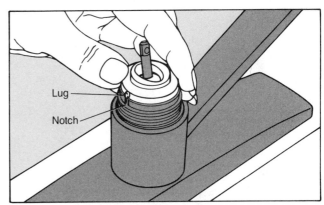

Lug

Notch

6 Next, replace the cam assembly. On the faucet body there is a small notch. The cam assembly contains a small protruding lug. Be sure to fit the lug into the positioning notch. Then replace the cap and screw it down securely.

CAUTION

Before working on faucets, turn off the water at the fixture valves or the main shutoff valve (page 9). Turn on the faucet and allow all water to run out.

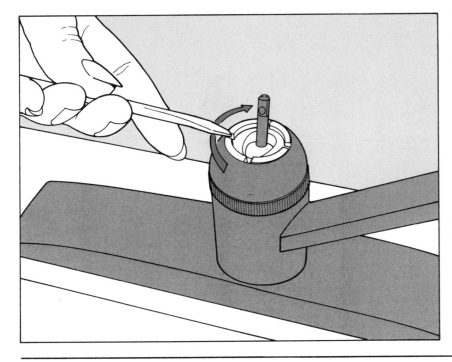

7 Move the stem of the ball to the ON position; water should not leak out around the stem. If it does, then you should tighten the adjusting ring by turning it clockwise with the tip of a small screwdriver. If you have to tighten it so tight to stop the leak that the handle is difficult to move, then the entire cam assembly, both the plastic part and the rubber ring, needs to be replaced.

Position the handle so that the setscrew is over the flat part of the stem; tighten the setscrew.

Repairing a Disc Faucet. Leaks from a disc faucet show up as a puddle around the base or under the sink. The problem is usually worn inlet seals or sediment buildup around the inlet holes. If seal replacement and cleaning of the holes does not cure the leak, then the discs are worn and you should try to replace them. Take all parts to the plumbing store for a duplicate. If you can't find matching parts, replace the entire faucet.

CAUTION

Before working on faucets, turn off the water at the fixture valves or the main shutoff valve (page 9). Turn on the faucet and allow all water to run out.

REPAIRING A DISC FAUCET

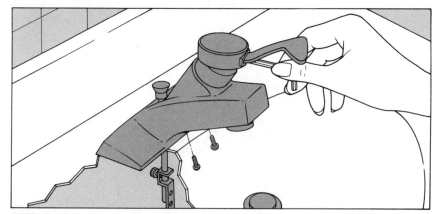

1 Disassemble this faucet by lifting up the handle as far as it will go to locate the setscrew; unscrew it to release the handle. Next, remove the chrome body cover. This is done by disconnecting the thumbscrew holding the pop-up drain rod beneath the sink. (Mark the rod first so that you know exactly where to reconnect it.) Unscrew the two Phillips-head screws on the underside of the faucet; lift off the body cover. On newer models, there is a setscrew in the handle and a chrome ring that unscrews; with these it is not necessary to remove the body cover.

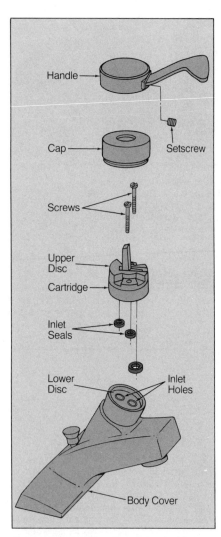

Handle

Cap — Setscrew

Screws

Upper Disc
Cartridge

Inlet Seals

Lower Disc — Inlet Holes

Body Cover

Anatomy of a Disc Faucet. In this type of faucet, two discs connect with the handle to mix hot and cold water. Usually one or both rubber inlet seals are worn and need replacement. A disc assembly rarely wears out.

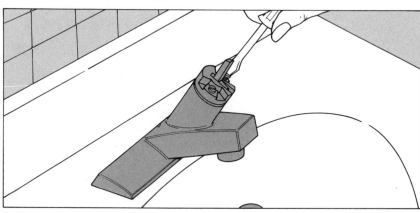

2 Remove the cartridge simply by loosening the screws that hold it to the faucet body. Under the cartridge are a set of inlet seals; remove them and replace any worn ones with exact duplicates. Also, check for sediment deposits around the inlet holes and scrape away any possible obstructions.

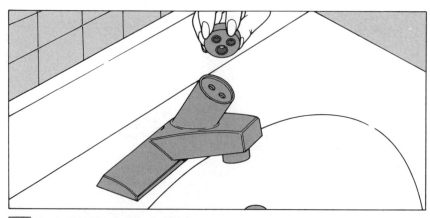

3 Replace the old cartridge (or install a new cartridge) by aligning the inlet holes of the cartridge with those in the base of the faucet. Reassemble the faucet by following step 1 in reverse.

Repairing a Tipping Valve Faucet. Though these faucets are no longer made, you can usually find parts for them, or kits with all parts, at plumbing supply stores. Leaks can be caused by several problems. A leak at the body is usually caused by a defective O-ring and spout drips are usually the result of a faulty valve seat or other valve part. Strainers sometimes clog up, causing the faucet to flow slower than usual and handles often become wobbly because of loose or defective screws.

CAUTION

Before working on faucets, turn off the water at the fixture valves or the main shutoff valve (page 9). Turn on the faucet and allow all water to run out.

REPAIRING A TIPPING VALVE FAUCET

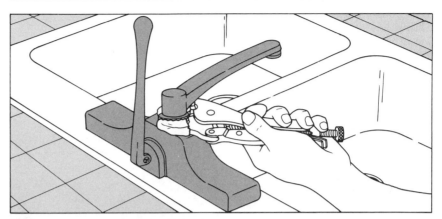

1 With a pipe wrench wrapped with tape, turn the spout ring counterclockwise until the spout can be lifted off. With a screwdriver, pry up the chrome body cover; lift it off. If the only problem is a leak at the base of the spout, replace the O-ring with a new, identical one; reassemble and test the faucet.

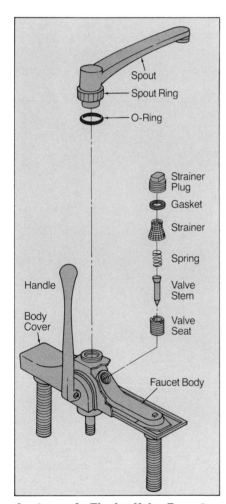

Anatomy of a Tipping Valve Faucet.
A tipping valve faucet has a pair of valve stem assemblies, hot and cold, through which water flows up and out of the spout. Moving the handle forward and backward controls the pressure; moving it from side to side controls the mix. All parts are replaceable.

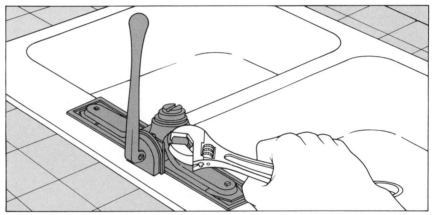

2 If the faucet has a leaky spout, you will have to replace a valve part or the valve seat. Continue to disassemble, unscrewing the strainer plugs on either side of the faucet. Remove the interior parts by hand, one by one—the gasket, strainer, spring, and valve stem. With a valve seat wrench, remove the valve seat. If the flow from the faucet has been sluggish, the strainers might be clogged; wash them in soapy water with an old toothbrush. Inspect all parts for wear and corrosion; replace any part that looks worn with a new, duplicate part. Lubricate the threads of the valve seat with plumber's grease before reinserting. Reassemble the faucet.

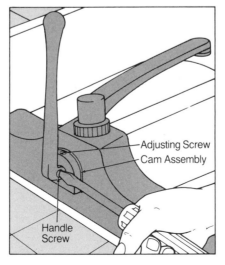

3 If the tipping valve faucet handle has been loose, tighten the handle screw that holds it to the cam assembly. If this doesn't solve the problem, remove the screw and inspect it. If the unthreaded part of the screw beneath the screwhead is badly worn, replace it with a new, identical screw. If the handle is still wobbly, tighten the screw above the cam assembly.

Repairing a Cartridge Faucet.
If your cartridge faucet leaks around the
body, you should first check to see if the
O-rings are damaged; replace them with
identical parts. If you make this replacement and the faucet continues to leak,
then you should check to make sure that
you have reassembled the faucet correctly and, if so, assume that the cartridge itself needs replacing. Be sure to
follow the manufacturer's instructions
when doing this step.

CAUTION

Before working on faucets, turn off the
water at the fixture valves or the main
shutoff valve (page 9). Turn on the
faucet and allow all water to run out.

REPAIRING A CARTRIDGE FAUCET

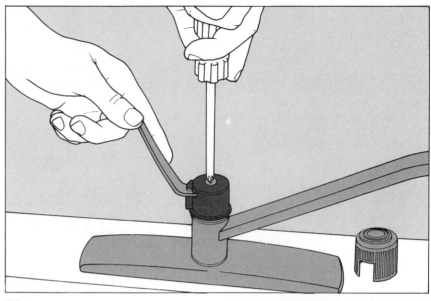

1 Disassemble the faucet by first pulling the handle cap off. Then remove the handle screw.

PLUMBER'S TIP: Not all cartridge faucets are designed like this. The handle screw might be hidden by a decorative button; if so, pry it off to expose the screw.

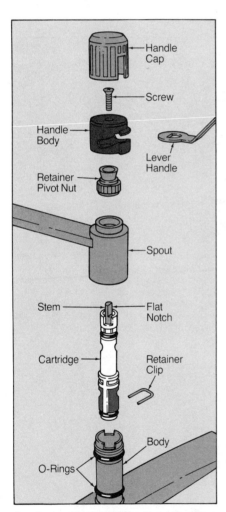

Anatomy of a Cartridge Faucet. A cartridge faucet has a series of holes in the stem-and-cartridge assembly that align to control the flow and the mixture of water. The cartridge is held in place by a retainer clip which is on the outside or inside of the faucet; remove it and the assembly lifts out easily. The most common problems require that the O-rings or the cartridge itself be replaced.

Labels (anatomy diagram): Handle Cap · Screw · Handle Body · Lever Handle · Retainer Pivot Nut · Spout · Stem · Flat Notch · Cartridge · Retainer Clip · Body · O-Rings

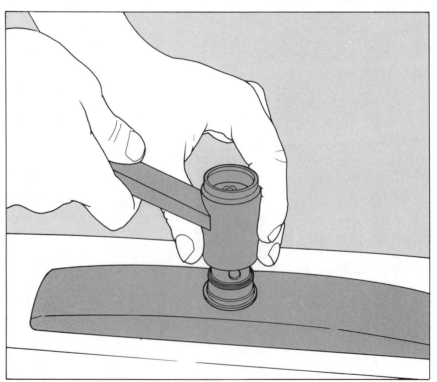

4 Twist and lift the spout off.

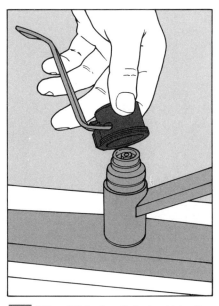

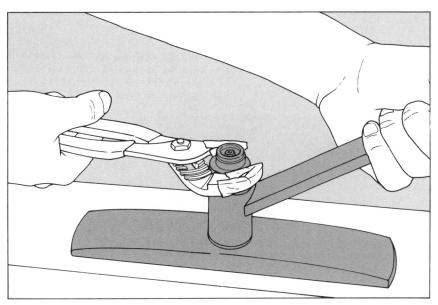

2 Lift and tilt the handle body which will contain the lever handle. Leave the handle inside the body.

3 With a tape-wrapped wrench, remove the retainer pivot nut.

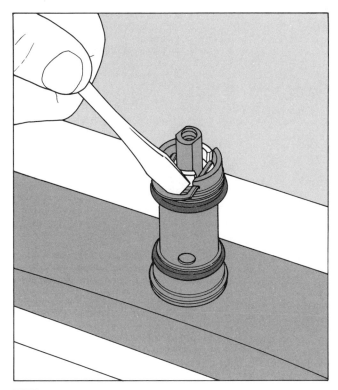

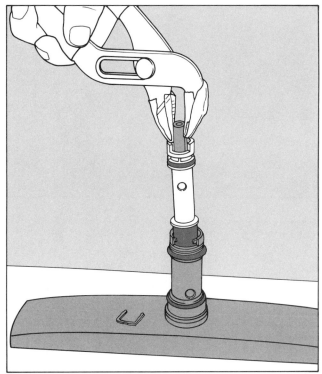

5 With a screwdriver, pry out the retainer clip. Grasp the cartridge stem with pliers and lift it out. Carefully check the O-rings for signs of wear and replace them if necessary. If your faucet is a swivel-mount type, first apply a coat of petroleum jelly to the O-rings.

6 If the O-rings look alright, replace the cartridge with an exact duplicate. This installation is not difficult but be sure to read the manufacturer's instructions. A common model has a flat sided stem that must be installed facing to the front; otherwise the hot and cold water will be reversed. Insert the retainer clip back in place, gently but snugly. Reassemble by putting the parts back together in reverse order.

Aerator, Spray Hose, and Diverter Repairs

These three components control the flow of water from the faucet and make it more convenient for our use. The most common problem with these 'flow fixtures' is that they become clogged with sediment buildup. The condition of your water supply is directly related to how often you will need to clean parts; you can lengthen their life by cleaning them regularly. When cleaning aerators, it is not necessary to shut off the water at the main; simply shut the faucets tight.

Though repair is much easier than that of faucets, there are some tips to follow. Aerator parts are very tiny and should be lined up and assembled precisely in order to function correctly.

Spray hose replacement is tricky only because of its position under a sink. The easiest way to remove it is to use a basin wrench (page 12).

Spray hose problems include leaks at the spray head and leaks at the base of the faucet spout. A simple cleaning of the spray hose aerator might be all that is warranted to put the system back into operation. Other places to check are the spray head where there might be a worn washer and the hose itself, which can be cracked due to wear.

A good cleaning is usually all that's needed to get a diverter valve back in operation. It's easy to disassemble, and the scrubbing can be done with an old toothbrush. Sometimes, however, cleaning won't suffice and you will have to replace it.

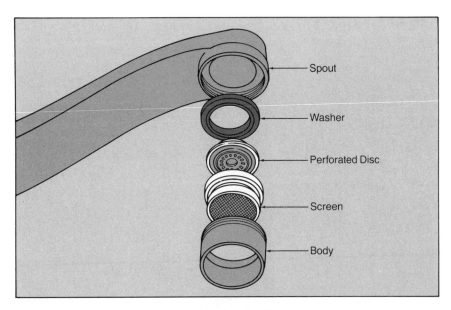

Anatomy of an Aerator. Most faucets have, at the tip of their spouts, aerators that mix air and water for a smooth flow. Aerators should be cleaned periodically to remove mineral deposits and debris buildup. Clean by disassembling and setting the parts aside, in order. Then use a brush and soapy water on the screens and disc. With a pin or toothpick, open up clogged holes in the disc. Replace any worn parts and then flush all parts with water before reassembling.

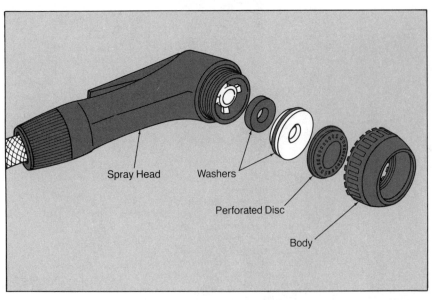

Anatomy of an Aerator for a Spray Hose Nozzle. Sink spray hoses also have aerators in their nozzles. When these are clogged, they can cause a diverter valve malfunction. Clean them using the same procedure as for cleaning an aerator (above).

REPAIRING A SPRAY HOSE

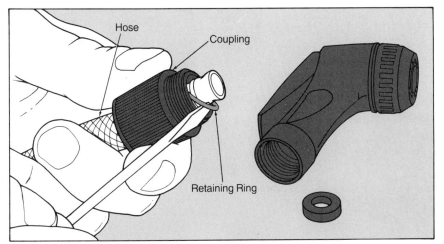

Hose
Coupling
Retaining Ring

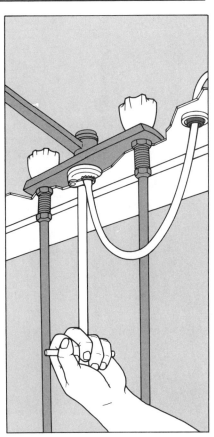

1 If the spray hose leaks at the spray head, unscrew it from the coupling at its base. Next, separate the hose and coupling by snapping off the retaining ring with a small screwdriver or penknife. If it is worn, replace the hose washer under the coupling. Then, to clear a blocked hose, run water full force through the hose. With the spray head removed, water will go through the hose rather than the sink spout.

CAUTION

Before working on a spray hose, turn off the water at the fixture valves or the main shutoff valve (page 9). Turn on the faucet and allow all water to run out.

2 If the hose leaks at the base of the faucet spout, unscrew the coupling under the sink. This is a single hex nut that is difficult to get to. You can try using locking pliers but the best tool for the job is a basin wrench, a special plumber's tool made for working in close quarters. Put on eye protection, lie on your back and work with a flashlight. Once removed, inspect the hose for cracks or kinks; replace a damaged hose. Purchase a replacement hose that has the exact same size diameter; nylon-reinforced vinyl works best. Also, take the hex nut with you to the plumbing store—you might need to purchase an adapter.

CLEANING OR REPLACING A DIVERTER VALVE

Diverter Valve

First, remove the faucet spout to get at the diverter (pages 62-70, depending on the type of faucet). Some spouts are secured to the body by a grooved ring. Chrome-plated fasteners are easily scarred so be sure to remove them with a tape-wrapped wrench. The tip of the diverter valve is usually capped by a brass screw. Turn it just enough to free the valve from the valve seat inside the faucet. Then pull out the screw and valve at the same time. Valves without screw tops can be pulled straight out with pliers.

The diverter valve is patterned with tiny inlet holes; these sometimes get clogged with mineral deposits. Clean the valve with a sharp but soft object such as a toothpick or with an old toothbrush. Do not use any metal objects for cleaning. Work at a different sink and periodically flush the valve with water running full force. Reassemble the faucet and test for performance. If it still doesn't work properly and you have corrected problems with the aerator and the spray hose, then purchase and install a new, identical diverter valve.

6

Planning a New Plumbing Installation

As you improve your home, the plumbing changes you may need can range all the way from adding another outside faucet to installing a large fixture or water-using appliance. Perhaps you've always dreamed of having a sink on your screened-in porch for potting plants, a wet bar in your recreation room, or a darkroom sink. Maybe you've discovered a better location for your clothes washer or wish to add another lavatory in a bathroom. This chapter will serve as a guide in deciding how much to do yourself and how to do it.

In Chapter 1, the basic plumbing system was explained and you learned how the supply, drain-waste, and vent pipes interacted with each other. In Chapter 3, you were introduced to the pipefitting skills essential for making new plumbing runs. Here, you'll put it all together and plan for a new plumbing installation. But before you can do this you need to take a closer look at the system. An understanding of traps and venting is essential.

The chapter will also give step-by-step instructions for a lavatory *rough-in*. A rough-in is the installation of all parts of a plumbing system which must be completed prior to the installation of a fixture. This includes DWV, supply pipes, and any necessary fixture supports. Lavatories are a popular installation and the how-to instructions will aid you when making other types of installations as well. (See the remaining chapters for other specific installations.) In addition, a wealth of general information is presented. Practical advice is given regarding materials, both type and size, and installation methods are discussed at length.

When the time comes to actually do the work, more than ever, safety is an issue. Plumbing runs can be very long and involved. If soldering, you will need to be constantly aware of the silently burning propane torch. If you're using solvents to join plastic pipe—make sure that your work area is well ventilated or that you take breaks often.

Planning the Project

The major considerations when planning a new piping run are: your local plumbing code, the limitations of your present system's layout, design considerations, and your own plumbing skills. All four of these factors must be carefully thought out and weighed for the plan to take shape.

Checking Codes

Practically any new pipe added to your system will require approval by your local building department. Obviously, the time to check is not after you have done the plumbing. Codes will specify not only what materials and methods you can use but also whether or not you can do the work at all. Some codes require that certain jobs be tackled only by a licensed plumber. One reason for

permits and inspections is insurance claims. For example, if an uninspected water heater were to explode, your insurance policy might not cover the damage caused by the explosion. Another issue for homeowners is the sale of the property. When the time comes to put your home on the market, you'll see that most realtors and potential buyers will be adamant about work that is up to code and inspected.

Simple extensions of supply pipes are rarely restricted, as long as your water pressure is suitable. Diameters of supply pipes serving fixtures or appliances, however, will be spelled out clearly. The code is definitely insistent about the DWV system; that you can be sure of.

Before even checking with the code, you should be prepared to answer the following questions:

■ Where can you place new fixtures within your present DWV system?

■ Is your present main stack, secondary stack, or branch drain adequate in size to accept new pipes?

■ By what method will the new fixture be vented?

■ Will your present water heater provide enough hot water for the new fixtures or appliances?

■ Is your water pressure sufficient for the extra demand of the new fixtures or appliances?

If you have trouble answering any of these questions, then your proposed installation plan is likely to have flaws and you'll have to go back to the drawing board and develop a new plan.

Mapping Out Your System

Any limitations in your system should become evident as you make a map of the present system. For background information, read pages 1-4. Then, start in your basement or crawlspace and make a sketch of your main soil stack, branch drains, the house drain, and

accessible cleanouts. Drain stacks are located within the *wet walls* of your home. Wet walls are often thicker than usual to accommodate large stack pipes.

Next, trace the network of hot and cold water supply pipes. Since supply lines often twist and bend in walls, it might help to turn on faucets and listen with your ear to the walls. Search until you find the pipe that is closest to your new installation and easily accessible. Finally, check the roof and attic for the route of the main stack and any secondary vent stacks.

When making the plan on paper, do it in this sequence: supply, drainage, and vent pipes. Decide what pipe materials you are going to use—either galvanized, copper, or plastic for supply pipes; plastic, copper or cast iron for DWV pipes. You can now sketch out your general plan. In doing so, try to be conservative in the placement of fixtures. Obviously, fixtures placed far apart from one another are going to cost you more in materials and work. (Note: Do not work out the *specific* plan until you have read everything in this chapter; at that time you can begin budgeting for your project—using the guide on page 124.)

Design Options

Planning a supply pipe run is usually no large problem. The DWV system, however, is more difficult to work with. Making a connection to the existing main soil stack is the simplest and the

most cost-efficient way to add a new fixture or group of fixtures. This can be done either individually or through a branch drain. Lavatories, tubs, or showers (but never toilets) can be tied directly into an existing branch drain.

A common approach is to make a tie-in for a new fixture or group of fixtures to the stack, either below or above an existing group. Check your code before assuming this to be acceptable. Another popular approach is to install fixtures back-to-back with a group of fixtures that is already attached to the main soil stack. If the fixture is a substantial distance from the stack, you'll probably need a 're-vent' pipe from the fixture's drainpipe to the vent stack above it.

Costs and work are accelerated when you want to make additions that are across the house from the existing plumbing. In this case, you'll probably need to run a new branch drain and a new secondary vent stack through the roof. Depending on what floor the installation is on and the design of your house, the new branch drain will tie into the vertical soil stack or the horizontal (sloping) main house drain via an existing cleanout.

All About Traps and Vents

After you read and thoroughly understand the basic plumbing system as explained on pages 1-9, it's time to take a closer look at the system. There are some rules about installations that

---

VENTING OPTIONS

Typically, homeowners have more than one choice in regards to venting. But before selecting one of these methods, check your local building code for restrictions.

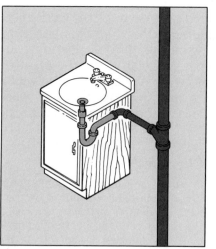

Wet Venting. This is the easiest method of venting. The fixture is vented directly through the soil stack or the branch drain.

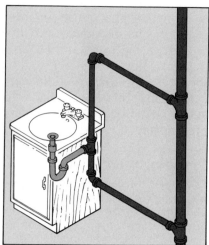

Back Venting. Also called *reventing,* this method involves running a loop up past fixtures to hook up with the main stack or the secondary vent; this connection must be above the fixture level.

aren't readily apparent either in this section or in the Skills chapter (pages 19-33). One of the things you'll need to understand if you're adding a fixture with a drainpipe is the principle of venting. And attendant to that is the subject of traps.

Sink traps were covered to some extent in the Repairs chapter (page 47), but what follows will give you a better view of the 'big picture' and how traps and vents function within the entire system. Building codes everywhere state that each fixture connected to a drain system must have a water-seal trap. The toilet itself is like one big trap. But our sinks, basins, and tubs have 'concealed' traps. Most commonly, a P trap is seen in home installations although a check of your system might reveal an S trap. The S trap, not shown here and no longer approved by most codes because it is not vented, ends in a vertical pipe so that water naturally flows downward. The P trap, in contrast, ends with a horizontal pipe. To aid the water flow this pipe must be pitched slightly downward.

Traps are one of the most important parts of a home plumbing system. They protect the health and welfare of the home's residents. For this reason, building codes are very tight in their regulations regarding traps. The safest practice, if you are replacing a fixture or adding fixtures similar to those already in your house, is to use a trap like that currently in use. On the other hand, if you're installing a new fixture unlike

what you already have, you should consult a building inspector or plumber for advice.

The Vent System. Venting is extremely important. Without proper venting you are liable to have sinks that drain too slowly or toilets that need to be flushed twice. Two other phenomena are caused by poor venting. *Negative pressure* siphons water from traps by suction, and *positive pressure,* (also called blow-back) actually forces sewage gas bubbles through the water in traps. Both of these force undesirable odors into the home.

There are, in fact, three kinds of venting procedures, all determined by local codes. They are: *wet, back, and individual* and they provide you with several options depending on whether vent pipes are available nearby or not. As always, you should choose the easiest method as long as your code permits it.

Extending the DWV System

This process is relatively easy with plastic or copper plumbing, and not so easy with cast iron or galvanized steel. A good example is adding an extra sink. The first step is charting the path the new drainpipe will take from the sink to the soil stack or waste stack. Depending on the situation, it may run from the fixture trap (directly under the fixture) along the wall under a counter, then down the floor and across the basement ceiling to the stack. This is only one possibility; there are countless other courses it may have to take, depending on the individual case. One factor to consider is that cleanouts are required on drains. Individual codes will vary as to the distance required between cleanouts. Check your code.

When the drain runs from one level of the house to another, as from the first floor to the basement, it's best to use vertical, rather than slanted piping. Short slanted sections, however, may be used to get around obstructions such as parts of the house framing. Lateral runs, as under counters or across the basement ceiling, should have a downward pitch toward the stack connection —from ¼ inch to ½ inch for each foot of run. The connecting fittings that join your vertical runs to your lateral runs provide for this pitch.

Unlike supply pipe fittings, drainpipe fittings are not always exactly what they seem. A 'right-angle' elbow, for example, is likely to form about a 92° angle instead of the true 90°. The extra 2° aren't apparent to the eye when you look at the fitting, but that's all it takes to provide the needed pitch. The same

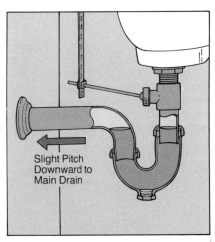

Understanding Trap Installations. This trap, called a 'P' trap because of its shape, works to hold or trap water and thereby prevent sewer gases from entering the home. The pipe directed to the main drain must be angled slightly downward to aid the flow of wastewater. Though not shown here, all traps must be vented.

applies to Ts that are used to connect a vertical vent line at some point along a lateral drain line. If the T isn't marked for the direction of flow, you can tell which way the branch slants by temporarily inserting lengths of pipe into the fitting. A few feet of pipe will reveal angles too small to discern in the fitting.

Adjust your plan so your new drainpipe enters the stack at a convenient location for a connecting fitting. This might involve varying the length of the vertical runs. Usually, it's possible to plan roughly with taut string and push pins or nails.

Stack, Drain, and Vent Sizes. The information regarding pipe sizing is available through your local code. It is determined by the number of *fixture units.* A fixture unit is a unit of measurement representing 7.5 gallons or 1 cubic foot of water per minute. All plumbing fixtures have fixture unit ratings listed in chart form in the code. Due to the general popularity of water conservation, there has been a recent trend toward slightly smaller drainpipes and also smaller vent pipes. This is evidenced by the latest NAHB (National Association of Home Builders) *Residential Plumbing Guidelines* which has been adopted by the CABO (Council of American Building Offices) one- and two-family dwelling code.

The soil stack into which your new drain line will empty is most likely to be 4-inch cast iron pipe. In many new houses, however, it may be 3-inch copper

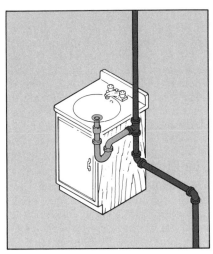

Individual Venting.   Also known as *secondary venting,* this involves running a new, or secondary, vent stack up through the roof. Used when there is a new fixture or group of fixtures far from the main stack.

tube. If it's copper, your connecting-in isn't difficult.

Drainpipe diameters vary by code, but if there's no code in your area you can use 1½-inch or 2-inch pipe for basin, sink, and tub drains. The smaller size cuts costs while the larger one speeds draining. Toilets connect to the soil stack with a 'closet bend', usually 4-inch diameter lead pipe.

Vent pipe sizes are based on the kind of fixture being vented, the diameter of the drainpipe, and the length of the vent pipe. Generally, the diameter of an individual vent can never be less than 1¼ inches or less than one-half the diameter of the drain that it serves, whichever is larger.

Extending the Supply System

This is usually an easy job for the homeowner. If the water-supply system is of copper tubing, the addition can often be completed in a matter of hours. It begins with the selection of a starting point for the new piping. In many instances you can remove an elbow fitting and replace it with a T fitting to provide the extra connection that starts the new run.

The first section of the new piping should be connected to the fitting at this time. If you have to do the job in stages because of limited spare time you can solder a cap on the end of the new run's first section (or any other point) so you can turn on the water and use your regular system until you have time to finish the addition.

If it isn't convenient to flex an existing joint apart to start your extension, you can remove a short section of straight pipe and replace it with another section that includes a T fitting with rigid tubing or flexible tubing. If it is to be led through walls or carried across ceilings or enclosed floors, soft tubing makes the work much simpler and reduces the chance of leaks.

Some plumbers prefer to use the rigid tube for neat appearance where plumbing is exposed and then switch to the soft tube for the unexposed runs. Remember, too, that you can place a valve somewhere along the line at an accessible point and keep it closed while you're working on subsequent stages of the job. This eliminates the trouble of removing temporary caps as you go along and it may come in handy for minor future repairs. If your code permits you to use plastic pipe instead of

copper, and you elect to do so, be sure to allow for 12-inch offsets in your design which should be made at 10-foot intervals.

Making Connections to Plumbing Fixtures

Fortunately, this is often the easiest part of the job. Both drain and supply pipes are connected to the fixtures by methods and fittings not used anywhere else in the plumbing system. The reason: the design of some fixtures makes the usual connecting methods impossible; and also, fixtures that may require replacement parts at times must be easy to disconnect. For example, traps (page 46) can be removed by screwing off both nuts and simply pulling down on the U portion.

Seeking Professional Help

The extension of piping requires the ability to accurately measure pipe runs, calculate DWV slope, and cut and join pipe and fittings. Along with these skills, general carpentry skills and tools are needed for opening up walls and floors and making pathways by notching or drilling framing members. Even trickier are bathtub or shower installations which usually require framing. Of course, the worst and messiest job of them all is adding a new soil stack or venting system.

If you have reservations about any of these tasks, you should hire a professional plumber to do the work. A viable option is to hire the plumber to approve and help you with plans and install the DWV pipes. You may or may not elect to do the supply piping yourself but making the fixture or appliance hookups should be within your domain.

If you want to learn about the roughing-in process, perhaps the plumber will let you be his assistant. On the negative side, plumbers, like all tradespeople, must protect themselves against lawsuits, so they might be reluctant to guarantee the entire installation if you have worked on part of it.

Getting Started

When making a new plumbing run, here are the basic steps involved: First, locate the existing supply lines and the stack that will serve as drain and vent. Then, connect the new piping to them and run the piping to the desired point. Finally, put on the new fixture. If your new plumbing will lead to a new room or to an unfinished room, install your pipes before the inside walls are put up.

GAINING ACCESS TO THE PIPING

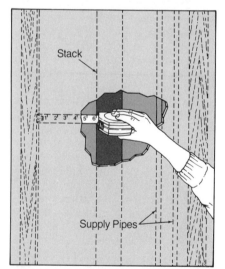

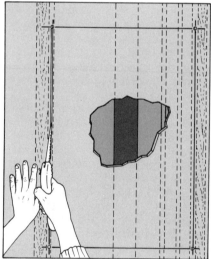

1 Wear eye protection for this project. Drill an exploratory hole to verify the location of the stack. Next, use a keyhole saw to cut a large enough hole into the wall so that you can measure for studs. Insert a steel tape measure into the wall and measure the distance to nearby studs. Determine the point at which you intend to tap into the stack. It is acceptable to use a point where another drainpipe enters. Mark the wall for a rectangular opening based on the measurements for the studs and 24 inches high.

2 Drill starter holes into the corners of the rectangle being careful not to drill into the studs. With a keyhole saw, cut into the wall at the edges of the studs. (Later, you can nail cleats to the studs to form a lip for a wall patch.) If this large opening does not reveal the needed supply pipes, locate them and repeat the process to gain access to them.

This is a lot easier than stringing it through afterward.

Flexible copper tubing or plastic pipes are the easiest and safest to use. Connections to existing piping can be made with adapters that link all kinds of materials. Any connections that cannot be reached later on without cutting into the wall should be soldered. These are likely to remain tight for the life of the house, whereas flare and compression connections may sometimes loosen slightly from vibration. The solderless connections, however, are very handy where you must work in cramped quarters or close to flammable building materials.

If you use soft tubing, chances are you'll need fittings only at the beginning and end of each pipe run. You might want to limit your soft tubing to vertical or near-vertical runs, or to lateral runs where enough pitch can be provided to clear all water from the wavy pipe when it is drained to prevent freezing.

If you must solder, which is likely, be sure to use a metal or other type of flame-proof sheeting for safety. Also, keep a fire extinguisher handy. Plumbers sometimes wet down an entire area where soldering will take place. For an easy pipe run, place pipes on outside walls and then conceal them with cabinets or other carpentry.

There's one thing that all codes insist on. A new drainpipe must enter the existing drain stack at a point low enough to make waste flow downhill, but not so low that the new trap will be sucked dry. This is accomplished by using the proper, maximum length of new drainpipe, called the *critical distance* (page 80). Fixtures do not necessarily need their own vent pipe; they may 'share' the vent of a nearby fixture. Generally this method is applicable only to lavatories; no more than three can share a vent pipe and they must be located close to one another.

Shown on the next several pages is a lavatory installation. Although the pipe run is shown on an outside wall, methods are also discussed for running pipe within the wall.

Finding Existing Lines

This step shouldn't be terribly difficult. Begin your investigation in your basement or crawlspace and then continue to map out the system on paper. Upstairs, listen for pipes through walls. When

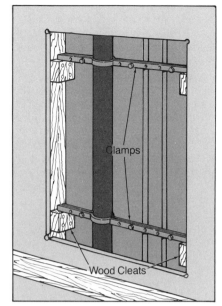

Preparing the Stack. A drain stack must be solidly anchored before you can cut into it. To do this, use stack clamps (steel clamps held together by bolts) above and below your intended working space. Position one strap behind the stack and one in front; tighten the bolts securely. To support the clamps, nail wood cleats into the studs, as shown. The cleats should be flush to the front edge of the studs. Leave the clamps in place after cutting into the stack.

MARKING FOR CONNECTIONS

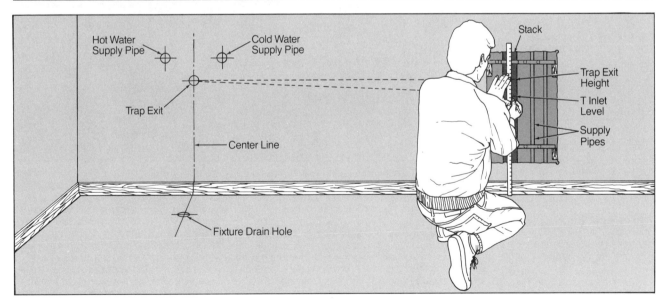

New fixtures usually include templates or other roughing-in guides to aid you in this step. For connecting a sanitary T to a stack, first check the critical distance so that you know the limitations of where your fixture can be located. Choose your preferred location and then on the wall, and floor if necessary, mark for the hot and cold water supply pipes, the trap exit, and the fixture drain hole. Through the mark for the trap exit, draw a vertical line. Measure from the center of the trap exit to the floor. Using this measurement, put a mark on the stack at the exact same height. Next, measure the horizontal distance from the trap exit to the mark on the stack. For each foot of this horizontal distance, subtract ¼ inch from the height that is marked on the stack. This figure will give you the shorter height—the exact location for your T inlet on the stack—and the correct slope that is necessary for drainage pipes. Measure for the locations of new Ts in the supply-line risers; since these pipes carry pressurized water they need not be sloped, and pipes can be positioned parallel to the floor.

If you are marking a back-to-back installation, reverse the steps and draw the line from the stack to the location of the new fixture.

you think that you have found their location, drill small exploratory holes. Avoid drilling into pipes or obstructions by stopping quickly when you meet resistance.

If you're reasonably sure that you have located pipes, then open up the wall for your work. Don't worry about cutting out a large section of the wall; if it's drywall, it will be no more difficult to replace than a small section. Cut out a rectangular piece from one stud to another.

After you have located the pipes, you need to anchor the drain stack for cutting. This is very important since even a minor slippage could break the seal at the top of the roof vent. Two clamps are required, one above and one below the tie-in (page 79).

Measuring and Marking

To establish the exact location of the new fixture, in this case a sink, you have to take several factors into account. The main consideration is the slope of the drainpipe for proper removal of waste. Since most codes recommend drainpipe measuring less than 3 inches in diameter, the slope will more than likely be ¼ inch per foot. Sloping drainpipes are limited in length because if they are too long they will empty a trap and render it ineffective. Be sure to use the correct critical distance as given in the following chart.

Determining the Critical Distance

Drainpipe Diameter	Maximum Distance to Stack
1½ inches	4½ feet
2 inches	5 feet
3 inches	6 feet

Although supply pipes don't need to be sloped, it is wise to follow this practice, as emptying them will be much easier. To determine the placement for the fixture pipes, you should use the fixture's template, usually provided with your purchase, to make your marks. Then measure backwards towards the stack, allowing for the slope. The illustration on page 79 shows an installation with a sanitary T.

If you're installing a sanitary cross for back-to-back fixtures, begin the measuring process at the cross and measure toward the fixture incorporating ¼ inch per foot. When you mount a new fixture next to an old one you must also use the ¼ inch per foot calculation, and the new fixture's drain hole must be within 6 inches vertically of the existing one.

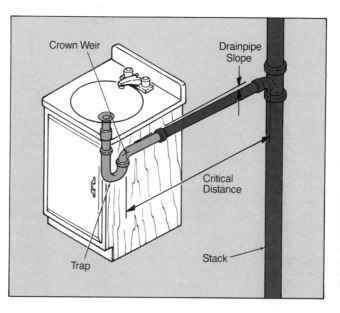

Crown Weir · Drainpipe Slope · Critical Distance · Trap · Stack

Understanding the Critical Distance. Drain outlets cannot be below the level of the trap's *crown weir;* if so, they would act as siphons to drain the trap. Accordingly, when the required slope of ¼ inch per foot is figured, the length of the drainpipe becomes limited. This limitation is the critical distance. Critical distances, set by local codes, also take into account the pipe diameter and the type of venting used.

MAKING SUPPLY PIPE CONNECTIONS

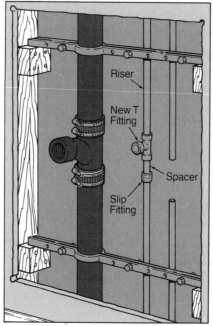

Riser · New T Fitting · Spacer · Slip Fitting

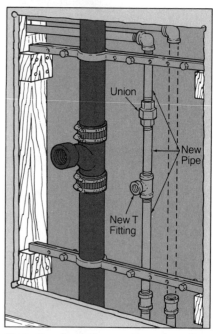

Union · New Pipe · New T Fitting

To Plastic or Copper Supply Pipes. Wear eye protection for this project. If the existing riser is made of flexible copper tubing or flexible plastic, cut out a section just large enough for the new fitting. Bend the tubing carefully to add the new T or cross. If the riser is rigid copper or plastic, cut out an 8-inch section of piping. Sweat a copper T or cement a plastic T onto the upper riser. Slide a slip fitting onto the lower riser and cut a spacer to fill the remaining space. While gently holding the lower riser to one side, cement or sweat the spacer to the T fitting. With the spacer and lower riser lined up, slide the slip fitting on and sweat or cement it into place. Depending on the play in the pipe, you might need to add an additional spacer or, for plastic pipe, install a spacer with one threaded end and a union.

For back-to-back installations, install new Ts above or below the existing Ts.

To Galvanized Pipes. Since galvanized pipes are threaded and cannot be unscrewed at one end without tightening the other end, you'll have to cut the pipes and follow them back to their nearest fittings at both ends. You may have to cut further into the wall to do this. Cut the pipe close to the fittings, and with two wrenches unscrew the remaining pipes. If you wish to install galvanized pipe, assemble new pipe, a T fitting, and a union, as shown. If you wish to change to copper or plastic pipe, add a transition fitting, a spacer, the T fitting, and new piping.

MAKING DRAIN-WASTE/VENT CONNECTIONS

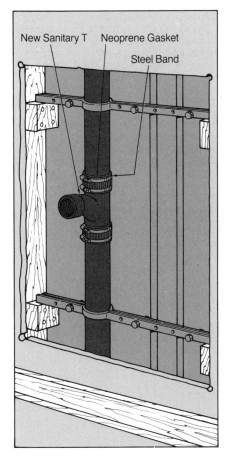

New Sanitary T   Neoprene Gasket

Steel Band

To a Cast Iron Stack. Wear eye protection for this project. This connection is made into a cast iron stack with a hubless fitting (page 16), a sanitary T. Hold the fitting at the stack and mark the stack for the top and bottom onto the stack, exactly where the fitting will enter. Support the stack as described on page 79; then cut into it ¼ inch above and ¼ inch below your marks with a special cast iron pipe cutter (page 31). Over both ends of the cut pipe, slip neoprene gaskets. Position the new fitting into place, slide the gaskets over the connections, and tighten the steel bands. It is not necessary to tighten all the way yet; this will allow you some flexibility when installing the connecting pipe. If you are going to run the pipe outside the wall as on page 83, angle the fitting 45° to the wall.

When installing back-to-back fixtures, you may need to cut out the existing sanitary T and install in its place a sanitary cross. The cross should be assembled between two lengths of hubless pipe used as spacers. The new cross should be at the level of the old T with inlets at right angles to the wall. This way the existing drainpipe can be easily screwed into one of the cross inlets.

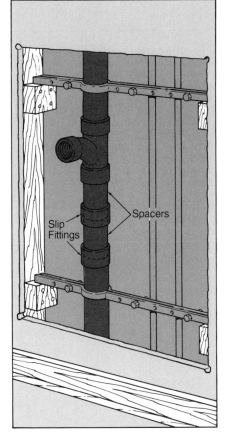

Slip Fittings   Spacers

To a Plastic or Copper Stack. Follow the marking and cutting directions as for cast iron pipe (above) and then solder or cement the fitting at the top of the cut stack angling it as needed for your run. Cut a short spacer, as shown, and join it to the bottom of the fitting. Measure for another spacer that will fill the remaining space exactly. Slip on two slip fittings, above and below the spacer, and then solder or cement them over the joints. Neoprene gaskets and steel bands may be used with a plastic stack as an alternative.

If you are connecting to a horizontal branch drainpipe, the procedure is the same but make sure that the pipe is properly supported with pipe hangers on each side of the cut.

CAUTION

Before making supply pipe connections, turn off the water supply at the main shut-off valve (page 9) and then drain the pipes, if possible.

Connecting the DWV Pipe at the Stack

Once you have made all of your measurements and marked the wall, floor, and pipes, you can begin making the connections. Tying into DWV lines entails cutting a section out of the stack and running pipe to a new fixture along pre-plotted lines.

Shown is a sanitary T connection but a sanitary cross can be installed if the project is for back-to-back fixtures. The most difficult part is cutting the pipe; fittings are fairly easy to put in. The process will vary depending on whether the stack is made of cast iron, copper, or plastic.

PLUMBER'S TIP: If a sanitary cross is used, a cleanout fitting should be installed above it. The reason: When using a snake, the snake will erroneously push through the cross rather than going into the vertical section.

Tapping the Supply Pipes

When tapping into copper or plastic risers, simply add a T. With galvanized pipe the procedure isn't so simple; you will have to remove the entire section of pipe by tracing it above and below to the nearest connections. This could involve tearing apart a large section of the wall. An alternative that might prove easier is to run supply pipe from some other fixture or part of the house. For example, if you ran pipe from the basement you could use compression fittings. These would save you time and also would be exposed in case of a leak, unlike concealed connections.

Running the New Pipes

Once you have connected to DWV and supply pipes, you can begin to run the pipe toward your fixture. Unless you run the pipes on the outside of the wall, you will have to run them through the joists and studs of your house. Check your code for any restrictions on this procedure and then select one of the methods on page 82.

Step-by-step instructions are also given for running all pipes outside of the wall (page 83). This method requires less carpentry and leaves less wall to be patched. For angling the pipes, use one 45° elbow for adding a T or sanitary T or two elbows for adding crosses. Temporarily support the pipes and then anchor them to the wall with pipe hangers.

RUNNING PIPES THROUGH JOISTS

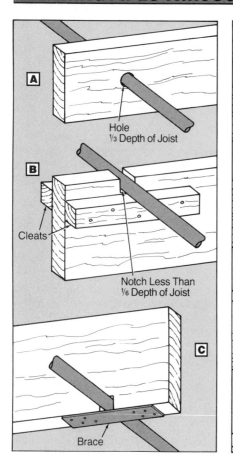

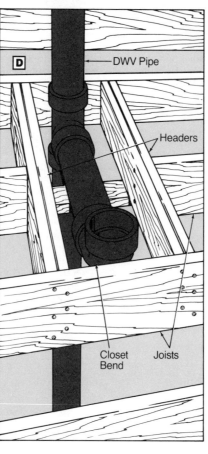

Depending on local code restrictions, you may use one of these methods when running pipes through joists. Be sure to wear eye protection when doing any drilling. When a pipe meets the center of a joist, you can drill a hole through the center for running the pipe but the diameter of the hole can be no greater than one-third the depth of the joist (A). Pipes may run through notches at the bottom or top of joists, but the notches may not be in the middle third (center) of the span and the notches can be no greater than one-sixth the depth of the joists. For pipe runs at the tops of joists, nail 2 x 2 wooden cleats as supports under the notch (B). For a joist notched at the bottom, use a steel brace or a wooden cleat as a pipe support (C).

Special provisions might be necessary when running drain-waste and vent pipe through joints. Depending on varying local codes, a joist may be cut into for a DWV pipe only at its end quarter. Once you have done this for the new pipe run, reinforce the cut by installing headers on both sides (D).

RUNNING PIPES THROUGH STUDS

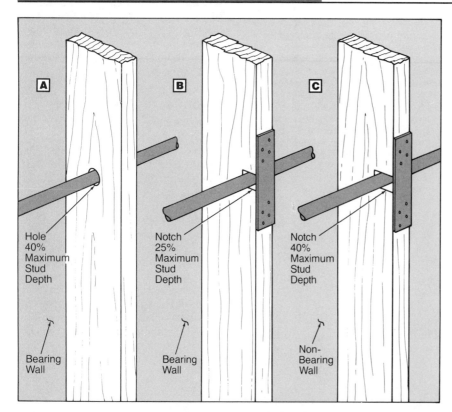

Depending on local code restrictions, you may use one of these methods when running pipes through studs. Be sure to wear eye protection when doing any drilling. For studs in bearing walls (those that hold up joists or rafters), you can drill center holes up to 40 percent the size of the stud depth (A). For studs in nonbearing walls (not shown), you can drill center holes up to 60 percent of the stud depth. When pipe runs to the edge of a bearing stud, you can notch the edge of it up to 25 percent of the stud depth (B). Nonbearing walls (C) can be notched up to 40 percent of the stud depth. All notches must be braced with steel ties.

RUNNING PIPE OUTSIDE A WALL

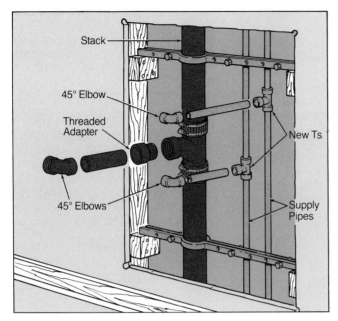

Stack

45° Elbow

Threaded Adapter

New Ts

45° Elbows

Supply Pipes

1 With the sanitary T and supply pipe T fittings angled at 45°, add pipes just long enough to extend outside the wall and to allow for new drywall. Onto the ends of these, slip 45° elbows. When connecting plastic drainpipe to a cast iron stack, use a threaded adapter, as shown. If you have installed a cross fitting at 90° to the wall, simply extend outside the wall with a long 90° elbow. As an option, install a pair of elbows, one at the cross inlet and the second beyond the wall surface.

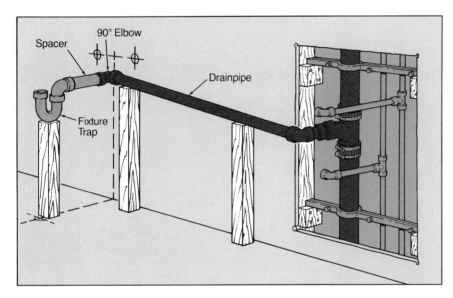

Spacer

90° Elbow

Drainpipe

Fixture Trap

2 Run the pipe along the wall beginning with the drainpipe. Assemble it loosely and prop the assembly up with scrap lumber or blocks and bricks. Position all parts directly at the marks made on the wall and floor. At the end of the drainpipe add a 90° elbow, a spacer, and the fixture trap. Be sure that the end of the trap meets your drain mark on the floor.

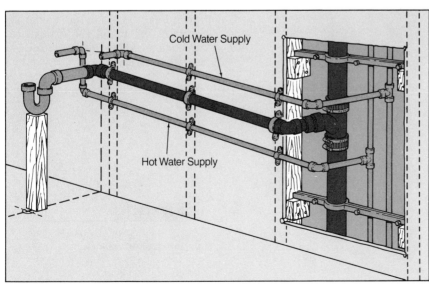

Cold Water Supply

Hot Water Supply

3 Sweat or cement the drain assembly and then anchor the pipe with metal straps to all the studs along its route. Repeat the entire process with the hot and cold supply pipes and add the necessary fittings to lead them to your marks on the wall. Add 90° elbows and short spacers. Install shutoff valves if preferred, install the fixture, and test the entire system for a few days before closing up the wall.

Finishing Up

Make the fixture hookup using manufacturer's instructions plus the appropriate instructions given in this book. (Lavatories are explained on pages 88 and 102; faucets on page 89.) Add shutoff valves for the supply pipes for convenience when working on the plumbing (right). Then, before replacing any wall covering that you have removed, test the new piping for leaks for several days. Patch the wall, floor, and ceiling.

If you have made the run on the outside of the wall, there are several methods that you can use to hide the pipes. You can thicken the wall with new studs and build a storage shelf or ledge above the pipes. For pipes on the floor, you can build a platform to mount the fixture on. Closets, shelves, and cabinets are also common concealing structures.

Adding Shutoff Valves

Shutoff or stop valves make life easier when problems arise. Small repairs can be made right at the fixture, without running up and down stairways to the main shutoff valve. Separate shutoffs also allow you to use your water elsewhere throughout the house during prolonged repair jobs.

On the other hand, such convenience might be unwarranted and time has proven that shutoff valves are prone to 'freeze shut' during long periods of disuse. Another reason for not adding them at every fixture is esthetics. If you have a decorative basin that has clean lines, you might not want to interfere with its good looks by adding shutoff valves, even if they do have chrome plating on them. Generally, local plumbing codes do not require the installation of separate valves for every fixture. However, fixtures for which they are recommended include: clothes washers, water heaters, dishwashers, and toilets.

Steps are given for installing valves to an existing fixture, an easy process. Of course, you may want to replace faulty existing valves, an even easier job. Obviously too, if you have just made a new piping run for a new fixture, then now is the time to install the shutoff valves. Although there are two kinds of valves, depending on whether the supply pipe comes from the wall or ceiling, only one is shown here, the angled type.

Supply pipe exits a wall by a short pipe called a *stubout* and it is here that the replacement or installation procedure begins. The instructions detail the addition of valves to a threaded wall stubout for a lavatory faucet but the process will be the same if you are adding straight or angled valves for kitchen sinks or toilets.

When you purchase your shutoff valve, be sure to take the existing connecting pipe or fitting from the stubout to the store with you; it's the best way to get a perfectly matched valve. Choose one that's compatible in material: with iron, use iron; with plastic, use plastic; and with copper tubing, use brass. If you don't get the proper type, be sure to first use a transition fitting before connecting the valve. If the valve will be visible, the chrome-plated type is recommended.

The next requirement is the flexible tubing that connects the valve to the fixture. Piping for other fixtures may need to be stronger, but for valves at sinks, toilets, tubs, or showers purchase a flexible tubing. These are available in different materials but the most popular type is a 3/8-inch chrome-plated copper tube made specifically for such connections. (The instructions presented here show its use.) These are available in 1-, 2-, or 3-foot lengths and may be simply cut to the exact length needed.

Kits are available that contain the valve already joined to the tubing; also included will be all the fittings and complete assembly instructions. Using the kit is especially easy; or you may wish to buy the separate parts and make the connections yourself (shown here).

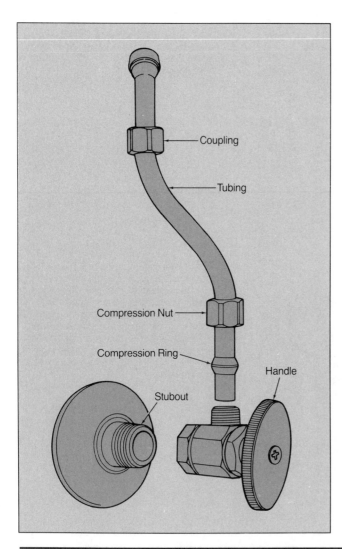

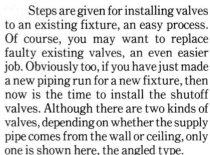

Anatomy of a Shutoff Valve. This shutoff valve includes all the basic components although you can choose either an angled valve, as shown here, or a straight valve. The angled valve connects to a stubout coming horizontally from a wall; a straight valve connects to a stubout rising from a floor. Gate valves (page 60) are available although globe valves are recommended.

Labels: Coupling, Tubing, Compression Nut, Compression Ring, Stubout, Handle

---

CAUTION

Before making a shutoff valve installation, turn off the water supply at the main shutoff valve (page 9) and then open the faucet to drain the pipes.

INSTALLING A SHUTOFF VALVE

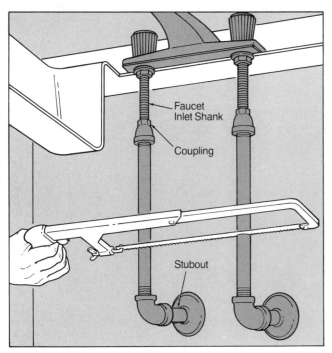

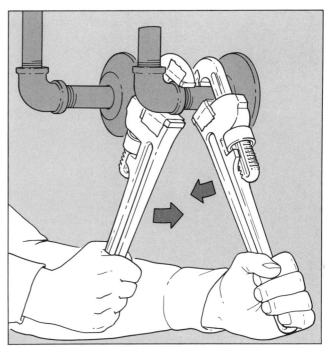

1 Cut a small ½-inch section out of the supply pipe close to the elbow. For galvanized pipe, use a hacksaw; for plastic or copper pipe, use a pipe cutter. Once cut, detach the shank coupling and remove the supply pipe.

2 Next, remove the elbow from the stubout. If galvanized, as shown, use two tape-wrapped wrenches. For copper tubing, unfasten the mechanical connection or melt the soldered joint. Plastic pipe must be cut. Detach everything except the stubout.

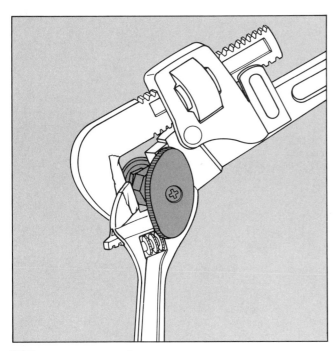

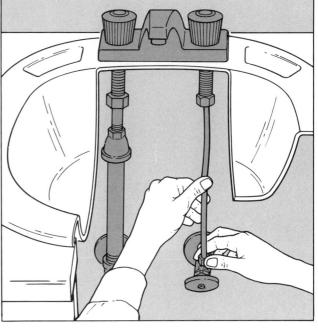

3 Prepare and clean the stubout to accept the appropriate fitting. For an unthreaded stubout, solder or cement a fitting to it. Screw in the shutoff valve after applying pipe joint compound to the fitting or pipe. Line up the valve outlet, as nearly as possible, directly under the fixture inlet. Hand tighten the connection; then continue to tighten with tape-wrapped wrenches.

4 Measure for and cut flexible tubing to run from the fixture inlet and inside the shutoff valve. With a basin wrench, fasten the coupling to the faucet inlet shank. Slide the compression nut and ring over the tubing's bottom end, insert it into the outlet, and tighten, first by hand and then with an adjustable wrench. Install a valve in the other line. Remove the faucet aerator, and run water briefly to clean debris from the pipes.

7

Plumbing in the Kitchen

Depending on your lifestyle, your kitchen could be the most active room of the house. Food is cleaned, prepared, and sometimes served here. Dishes, pots, and pans all pile up to no avail. Then the avalanche is cleaned and cleared and it's time to start all over again! In this chapter all the modern plumbing fixtures and appliances found in this much-used room are addressed.

Kitchen faucets and spray hoses need repairing relatively often (pages 59-73) and then, at some point, they need replacing. Instructions are given for installing a new deck-mount faucet and though this project takes place in the kitchen, the same basic process can be applied to mounting new lavatory faucets. Kitchen sinks get battered. Stainless steel gets scratched and stained; porcelain gets chipped. Since so much time is spent at this particular basin, it makes a wonderful home improvement to replace an old worn sink with a shiny new one. Details are given for this project as well as a brief account of all sinks—so that you'll understand their structural differences.

Step-by-step guides are given for two common kitchen installations—a garbage disposer and a dishwasher. When you have these new appliances professionally installed, there is often the aggravating realization that a serviceman has just spent 15 minutes in your home doing what might have taken you longer to do, but charging what it takes an hour or two for you to earn! You should use judgement about whether or not to install them yourself; read the material here but also read and follow the manufacturer's directions carefully. Once you have installed your new disposer, follow the advice presented here on its proper use. Also, refer to the troubleshooting guide if you need to fix your dishwasher.

As usual, check your local codes before doing any work. Also, be sure to follow the safety procedures here and in manufacturers' manuals; this is critical when doing plumbing that involves electrical wiring.

Modernizing with a New Deck-Mount Faucet

If your attempts at repairing an old kitchen faucet have proven fruitless or you've reached a point where the replacement parts are costly, this can be the impetus for replacing the entire fixture. Especially if you still have the old two-handled type faucet, you should consider supplying your kitchen with the standard convenience—a single-lever type faucet. Not only will it work better, but its modern appearance will be satisfying. The replacement process, as you'll see, is no more difficult, in most cases, than repairing the old faucet.

When you shop at the plumbing supply store for your new faucet, you'll find a myriad of choices. As mentioned on page 60, single-lever (washerless) faucets come in four types: ball, disc, cartridge, and tipping valve. You'll have to decide among these, plus you'll also have a selection of many different designs—everything from reproductions of antiques to unique futuristic creations.

The most important step in making your choice is to get the correct size faucet. The surest and easiest way to do this is to take the old faucet to the store with you. If this isn't possible, then you must measure the distance from the center of one inlet shank to the other. (The inlet shanks are the holes in the fixture for the tailpiece connections.) Also, you should measure the supply pipe diameters.

Check the amount and size of holes in the sink and purchase a faucet that will fit them. If you wish to add a faucet with a spray hose, this can be done without any additional plumbing, as long as you have the correct amount of holes in the sink. Faucet sizes currently available include those with 4-inch, 6-inch, or 8-inch separations.

If you cannot find a single-lever faucet to fit your sink, the options are to make the replacement with individual faucets or to replace the sink or lavatory. Do not attempt to drill holes into a fixture. If you need to cover any unused holes, however, this can be done easily with chrome escutcheon plates made for this purpose.

Besides size and style, another factor to consider in your purchase decision is quality. Generally, buy the best you can afford, within reason, and look for detailed installation and repair instructions. Kits are also available with

replacement parts, a real boon when maintenance-time comes.

Shown at right are steps for installing a new kitchen deck-mount faucet with a spray hose, but additional reference will be made to a bathroom lavatory installation. Remove your old faucet with the aid of a basin wrench and simply bend the tubes to fit into the center hole. Faucets are sealed with rubber gaskets or plumber's putty.

A Short Lesson About Sinks

Sinks and lavatories are classified according to how they are positioned within the house structure. In the next chapter on bathrooms (pages 100-102), there is a pedestal-type lavatory that is sealed to the floor. There are also several types of wall-mounted lavatories that are attached to walls by various clamps and brackets.

Deck-mount sinks are fixtures that are mounted into or onto a base, usually a countertop. As for kitchen sinks, deck-mount sinks are the standard. There are three different types depending on

how they are mounted: *self-rimmed, unrimmed,* and *frame-rimmed.* Among the three types, the most popular for kitchen use is the self-rimmed type although the frame-rimmed type is also used. Unrimmed sinks are most generally seen in bathrooms.

Installing a New Kitchen Sink

Kitchen sinks get their share of stains and abuse, but before you decide to replace the entire unit, consider minor repairs. If your porcelain sink has small chips, you can touch up the problem areas with matching enamel paint available at hardware stores. Leaky or defective strainers can be replaced by following the guide on page 44. On the other hand, when your kitchen sink has 'had it', or if you simply wish to redecorate—a new fixture is bound to brighten your culinary life.

Removal of the old sink can be very easy or very messy, depending on how it is set on the counter. If the sink is frame-rimmed or if it is a self-rimmed type set

INSTALLING A NEW KITCHEN SINK

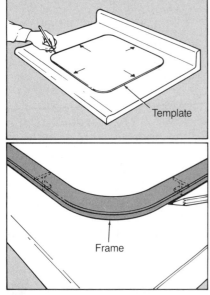

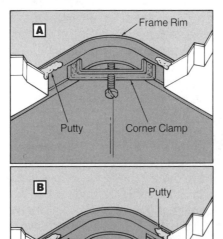

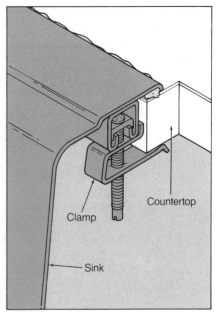

1 If you are making a replacement, shut off the water supply (page 9), drain and disconnect the supply pipes (page 89) and the sink trap (page 46). If there are clamps or lugs below the sink, remove them and then force the sink free.

For a new installation, mark the exact location of where the sink will be. Either use a template or, for a frame-rimmed sink, use the frame and mark the bottom edge of it. With a sabre saw, cut an opening in the countertop.

2 Attach the strainer to the sink using the basic directions on page 44 (but in this case the sink will be free from the countertop). Attach the locknut and tailpiece and then begin the framing process. If your sink is frame-rimmed, apply a coat of plumber's putty around the sink's top edge. Follow the manufacturer's instructions for attaching the frame to the sink. A common method (A) involves using metal corner clamps or lugs. Another method (B) specifies bending metal extension tabs around the sink lip.

3 A self-rimmed sink or an assembled frame-rimmed sink is set into the countertop. First, apply a ½-inch-wide strip of putty to the edge of the countertop. An option is to use silicone adhesive, but keep in mind that replacing the sink might be needlessly difficult if it is set too securely. Insert the sink into the hole and press down at all edges. Remove the excess putty and use clamps to anchor the sink at intervals, approximately 6 to 8 inches apart. Hook up the faucet (top, right) and trap (page 46); then turn the water back on and check for leaks.

INSTALLING A NEW DECK-MOUNT FAUCET

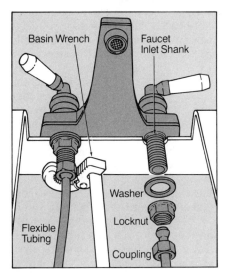

Basin Wrench
Faucet Inlet Shank
Washer
Locknut
Flexible Tubing
Coupling

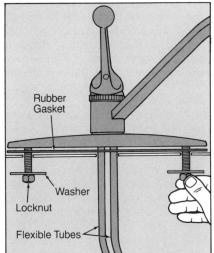

Rubber Gasket
Washer
Locknut
Flexible Tubes

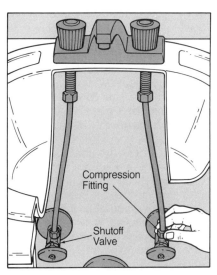

Compression Fitting
Shutoff Valve

1 First, remove the old faucet. Wear eye protection and use a trouble light with a bulb guard. With a basin wrench, remove the couplings that connect the flexible tubing to the faucet inlet shanks. (If you are working on a bathroom lavatory, you will remove the pop-up next, page 45.) Also with the basin wrench, remove the locknuts and washers from the shanks. If the sink has a spray hose, detach the coupling that connects the hose to the hose nipple under the faucet body (page 73, step 2). Lift out the faucet. Remove the flexible tubing by unscrewing a threaded joint, using an adjustable wrench on a mechanical connection or melting solder on a sweated joint. If you use the sweating method, be sure to take precautions against fire and burns.

2 Prepare for the installation by straightening the flexible copper tubes so that they will both fit into the center hole. Most faucets come with rubber gaskets on their bottoms; if yours doesn't, set the faucet on its side and apply plumber's putty to the bottom edges and at fixture holes. Clean off the surface of the sink. If the sink has a spray hose, install it first by slipping the hose down through its hole in the faucet, its hole in the sink, and then up through the sink's center hole. With a small adjustable wrench, attach the spray hose nut to the supply stub. Set the faucet in position while feeding the flexible tubing through the middle hole and then press it onto the sink's surface. Insert washers and locknuts onto the faucet inlet shanks by hand, underneath the sink. Tighten them with a wrench.

3 Bend the flexible tubing to line up with the connections at the shutoff valves. If the sink does not have shutoff valves or if you need to make extensions, use the methods on pages 19-33. Join the tubing to connections using compression or flared fittings. If necessary, use adapters to fit threaded parts to unthreaded tubing. Tighten all connections. If you have already assembled the aerator to the spout of the faucet, remove it. Run both hot and cold water for a few minutes to clean out the pipes and to check for leaks.

CAUTION

When installing a new faucet, turn off the water at the fixture shutoff valves or the main shutoff valve (page 9). Then open the faucet to drain the pipes.

THREE TYPES OF DECK-MOUNT SINKS

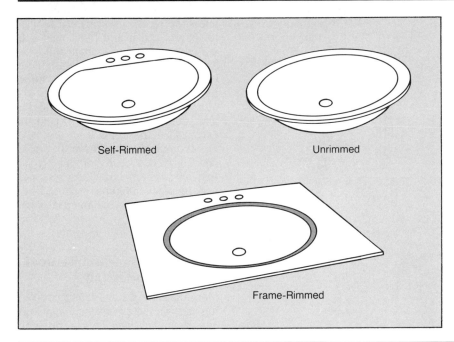

Self-Rimmed

Unrimmed

Frame-Rimmed

These three kinds of sinks require different installation methods although they are all sealed onto a countertop with clamps or lugs and plumber's putty. A self-rimmed sink has a molded overlap that rests on the edge of the countertop cutout. The unrimmed sink is recessed beneath the countertop and is held in place by metal clips. The frame-rimmed sink includes a metal strip that secures the sink to the countertop. The most popular type for kitchen use is the self-rimmed sink.

on top of tile or plastic laminate, the job will be fairly simple. Trouble arises when the sink has been set into a plywood cabinet and ceramic tile has been placed over its rim. In this case, you will have to remove all of the tile that rims the sink or, in many cases, remove the entire countertop. Wear eye protection and gloves. If you do dismantle the countertop, don't make the same mistake again; install the countertop first, then the sink.

Instructions are given on page 88 for replacing a kitchen sink and for making a brand-new installation. Frame-rimmed and self-rimmed sinks are addressed and although a kitchen sink is shown, the same steps can be applied to a deck-mount bathroom lavatory. You might want to install the faucet *first,* before the sink.

Away with Waste— Garbage Disposal Hookups

When this small garbage-grinding machine was widely adopted by homeowners in the 1960's, its use caught on quickly. Since then technology has provided us with better and stronger models to choose from. Controversy has developed, however, surrounding the use of garbage disposals. Certain geographical areas prohibit their use—and for good reason. The sewage disposal facilities in these highly populated areas simply can't tolerate garbage-laden wastewater.

Ironically, other communities actually encourage the use of garbage

disposals because their wastewater facilities are adequate and solid waste collection services are consequently kept to a minimum. So the first thing that you should do before considering a disposal installation is to check with your local building department. While you're seeking approval for the installation, also check to see what restrictions exist for the hookup.

Some codes require a separate trap and drainpipe when you have a two-compartment sink, whereas others permit you to hook the disposal up to a single drainpipe that serves both basins. Another common arrangement below the kitchen sink involves hooking up the dishwasher drainpipe to the top of the garbage disposal. This method is easier than tying in to the trap drainpipe and most garbage disposals are designed with a knockout for the hookup.

Once you have checked with your local code, purchase your new disposal. Check consumer guides for models that are trouble-free and also consider how you intend to use it. Some models might restrict you to grinding up softer materials than others. Additional features include reset buttons and automatic shutoffs that protect both the appliance and the house wiring if jamming occurs.

Make sure that installation instructions are clear, especially concerning electrical hookup. Disposals require a 120-volt grounded outlet and a drain outlet of not less than 1½ inches. If you already have the needed electrical outlet, you might consider buying a plug-in

type model. Otherwise, the fixture will have to be wired directly to your home wiring. The electrical work involved can be very dangerous, so if you have never done wiring before, you should hire an electrician to do this step.

As for the plumbing, you might have to change your sink trap to accommodate the garbage disposal (see page 46 for trap information). Otherwise it is similar to installing a sink strainer (page 44). Replacing an old disposal should be relatively easy since the connections are already available. When installing a new one, follow the installation instructions here and also use those that are provided by the manufacturer.

Garbage Disposal Maintenance. Once you have installed your new disposal, there are a few steps you should take to ensure a minimum of problems and a long life for it. This checklist applies to almost all brands.

■ Always use cold water when running your garbage disposal. Allow the water to run the entire time that the disposal is on and then let it run even longer to flush away congealed grease and other waste.

■ Avoid overloading; instead, grind up moderate amounts gradually.

■ Never dispose of the following items: cans, bottle caps, glass, rubber, plastic, foam, rags, string, paper, cigarette filters, or any similar materials.

■ Follow the manufacturer's instructions precisely about what the disposal will accept. Even though some models claim to handle seafood shells, bones and corncobs, use caution. If trash disposal is not a big problem, throw such 'hard' waste into the trashcan.

■ Deodorize and clean the unit in one of the following ways: Fill it ¾ full of ice cubes; run it and flush with cold water, then add ½ of a lemon and grind again. Or, flush the disposal with ½ cup of washing soda (available in supermarkets) which has been dissolved in a quart of hot water. Protect eyes and skin when mixing the solution and pour the excess down the drain.

If your garbage disposal breaks down or jams, switch it off quickly and then follow the manufacturer's instructions for unjamming and resetting. Don't attempt to use any commercial drain chemicals and, most importantly, *don't ever put your hand in a disposal.*

Dishwashers—Mechanical Hands in the Kitchen

More and more, dishwashers are becoming as essential a household appliance as a vacuum cleaner. This is due partly

Garbage Disposal Hookup To a Double Sink. Often plumbing codes permit you to hook up a disposal to a double sink as illustrated here. The wastewater is lead by a horizontal pipe to the trap on the adjoining sink. A directional T is required at this connection.

Directional T

INSTALLING A GARBAGE DISPOSAL

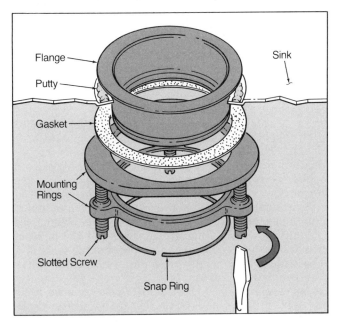

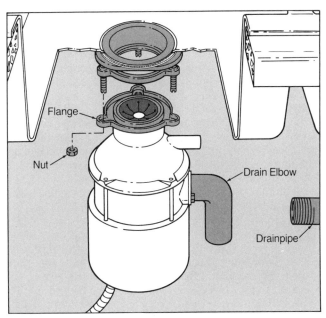

1 Remove the tailpiece and the trap from the sink strainer (page 47). Disassemble the sink strainer and pull it out of the sink (page 44). Thoroughly clean and wipe away old putty or sealing gaskets from around the opening. Apply plumber's putty to the sink opening and then seat the new sink flange in it. Put on eye protection and use a trouble light with a bulb guard. From below the sink, slip the gasket, the mounting rings, and the snap ring onto the neck of the sink flange. The snap ring fits into a groove on the flange to hold things together temporarily. One by one, tighten the slotted screws in the bottom mounting ring. Tighten until the gasket is snug and then remove any excess putty in the sink.

2 Hook up the drain elbow to the disposal. Next, align the holes in the flange of the disposal with the slotted screws in the mounting rings. Turn the disposal until the drain elbow lines up with the drainpipe. Tighten the nuts onto the slotted screws for a secure seal.

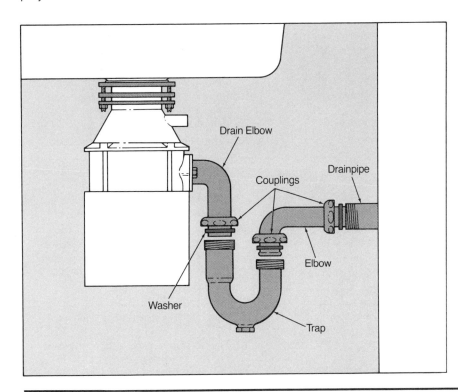

3 Hook up the drain by fitting a coupling and washer onto the drain elbow. Then connect the trap to the drain elbow fitting. Onto the other end of the trap, add an elbow fitting. You might need to cut the elbow to make the connection to the drainpipe. Test for leaks by running water through the disposal and tighten loose connections if necessary. You are now ready to make the electrical connection.

WARNING

For any installation involving both electrical and plumbing systems, use extra caution. Make sure that the electrical power is shut down at the circuit breaker or fuse box and follow the manufacturer's instructions precisely for the electrical hookup. Do not turn the power back on until all connections, especially ground, are made correctly and securely. If you have never done wiring before, it's recommended that you hire an electrician.

to their improved designs and partly to the change in kitchen design to accommodate their use. The most popular type is the built-in dishwasher which usually slides under a countertop or includes a countertop; regardless, the built-in type has its plumbing hooked up to your home's plumbing system.

In contrast, there are portable models available which require no plumbing at all. The water supply is drawn from a kitchen faucet and the drainpipe is simply hung over the kitchen sink. Portable models that can eventually be converted to built-ins are also available.

Local codes rarely restrict the use of dishwashers but they sometimes have tight regulations about the installation of built-in models. If your home already has a dishwasher, you can generally make a replacement without checking the code. But if you're making a first-time installation, you should definitely call your local building department. The installation guidelines might be stringent and a permit and inspection might be required.

Dishwasher Installation Procedures. Like a garbage disposal, the dishwasher also requires both plumbing and electrical work. One of the things that you'll need to prepare is a grounded junction box for directly wiring the appliance. An option is to use a grounding receptacle for acceptance of the machine's 3-prong plug. If you have never done wiring before, it's recommended that you call an electrician to do this part of the job.

Standard size dishwashers are made to fit into standard size cabinetry. The opening, therefore, should measure a minimum of 24 inches wide, 24 inches deep and 34½ inches high. It should be square and plumb and the floor beneath should be level; all this is necessary for the machine to work properly.

The three basic steps in plumbing for a dishwasher are the supply pipe connection, the drainpipe fitting, and the venting hookup. Since dishwashers perform their function with the use of hot water, only one supply pipe needs to be hooked up. This run is made with flexible tubing from the sink supply pipe. Along it, you should install a shut-off valve. The dishwasher's drainpipe is an easy-to-attach rubber hose which you will connect to the kitchen sink trap or to your garbage disposal, depending on the situation.

If a dishwasher has a pump to discharge wastewater, it must also have a drain line that rises to the total height of the washer. This high loop must be free from crimps to do its work of preventing wastewater from entering the dishwasher in case the kitchen sink becomes clogged. Some codes require, instead of the loop, the installation of an air gap. This device is easily added since it too has rubber hoses and clamps. The air gap must either rise above the countertop or, more commonly, will be installed in one of the kitchen sink knockouts.

Dishwasher Maintenance. If your dishwasher isn't terribly old and you suspect that a problem is in the plumbing of the machine, often you can fix it yourself. On the other hand, if you've eliminated all of the possibilities of faulty plumbing and the problem appears to be mechanical, such as a faulty solenoid, you should call a repairman.

Dishwashers have common parts and you should have no problem identifying them, especially if you refer to maintenance diagrams supplied by the manufacturer. Use this information and the troubleshooting guide (below) to diagnose your dishwasher malfunction and make minor repairs. Remember to use caution. Unplug the machine, shut down the power to the circuit, or shut off all power in the house.

TROUBLESHOOTING DISHWASHER PROBLEMS

TROUBLE	POSSIBLE CAUSE	SOLUTION
Dishes are not clean.	Water pressure is too low.	Call your water company.
	Water temperature is too low.	Adjust the temperature at your water heater to 120-130°.
	Spray arm is jammed.	Check for any obstructions at the spray arm.
	Spray arm holes are clogged.	Clean the residue from the spray arm holes.
Dishwasher won't fill enough.	Water inlet valve is closed.	Open the water inlet valve on the supply line.
	Water inlet screen is blocked.	Clean residue off the screen. Replace the water inlet valve.
	Water pressure is too low.	Call your water company.
	Solenoid is faulty.	Have the solenoid repaired or replaced.
Dishwasher fills too much.	Float switch is faulty.	Check it by jiggling; if it doesn't work freely, replace it.
	Dirty water inlet valve.	Disassemble and clean the valve parts.
	Solenoid is faulty.	Have the solenoid repaired or replaced.
Dishwasher doesn't drain properly.	Strainer basket is clogged.	Clean grease, dirt, and food particles from the strainer.
	Drain is clogged.	Unclog the trap cleanout under the sink.
	Air gap is dirty.	Clean air gap with a wire.
Dishwasher is leaking.	Hose connection is faulty.	Turn off the electricity; then tighten or replace hose.
	Door gasket is faulty.	Tighten or replace a loose gasket; replace a torn or split gasket.

INSTALLING A DISHWASHER

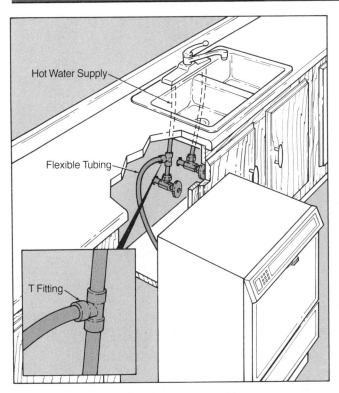

Hot Water Supply

Flexible Tubing

T Fitting

1 Install a T fitting into the hot water supply pipe, as shown, using the directions on page 80. As an option, you can install a three-way valve. From the T, run flexible tubing to the dishwasher. If preferred, along this run, install a shutoff valve (page 84).

CAUTION

Before installing a dishwasher, turn off the water supply at the main shutoff valve (page 9) and then drain the hot water pipe that you will be tapping into.

WARNING

For any installation involving both electrical and plumbing systems, use extra caution. Make sure that the electrical power is shut down at the circuit breaker or fuse box and follow the manufacturer's instructions precisely for the electrical hookup. Do not turn the power back on until all connections, especially ground, are made correctly and securely. If you have never done wiring before, it's recommended that you hire an electrician.

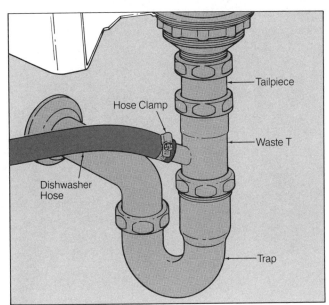

Hose Clamp

Tailpiece

Waste T

Dishwasher Hose

Trap

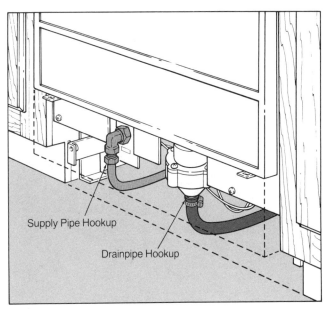

Supply Pipe Hookup

Drainpipe Hookup

2 If your dishwasher will drain into the sink trap, you need to install a threaded waste T fitting into the trap. First, remove the sink tailpiece (page 47) and insert the fitting into the trap. Tighten the coupling to secure it to the trap. Cut the tailpiece to fit between the waste T and the sink strainer. Reattach the tailpiece and then attach the dishwasher hose to the waste T with the hose clamp.

If your dishwasher will drain into the garbage disposal (page 90), begin the installation by shutting down the circuit that the disposal is on. With a screwdriver, punch out the knockout plug inside the disposal for this connection. With a clamp, attach the dishwasher drain hose to the drain fitting.

3 Depending on your local code and on the manufacturer's instructions, you must next vent the dishwasher to prevent a backup of wastewater. This is sometimes done by making a loop with the drain hose to the full height of the machine; if you use this method be sure not to crimp the hose. Another less common method requires that you add an air gap. Once vented, slide the dishwasher into place and make the hookups at the bottom of the machine for the drain and supply pipes. Follow the manufacturer's instructions for this step. Level the machine and anchor it to the countertop. Follow manufacturer's directions or call an electrician to make the electrical hookup. Finally, turn the water back on and check for leaks.

Bathroom Improvements

Years ago, the traditional three-bedroom home had one bathroom to be shared by everyone. How times have changed! Today's typical suburban home has a master bathroom, a 'hallway bathroom' for guests, and often another half-bath (toilet and basin) near the utility room or garage. Consequently, there's no more stacking up at the bathroom door on those tense weekday mornings when everybody's running late.

This chapter will take you all the way from removing your existing fixtures and replacing them with new ones to planning for a brand-new room. Aside from decorating, there are, of course, other reasons to install a new sink or tub. If your repair attempts (Chapters 4 and 5) continue to fail or you can no longer purchase parts for repair, then you might be forced to make a replacement. Or maybe you have one of those 'borderline' fixtures—it's not really broken but that small chip or stain is an annoyance to the eye.

Fixture designs and materials have changed through the years and these can provide incentives for making replacements. A new water-conserving toilet will help you to save on your water bill. Bathtubs are now made with pop-up stoppers, and new hand-held showers allow you more flexibility when bathing. Fiberglass, because of its light weight, makes a shower stall installation much simpler than ever before. Many of these replacements are easy and they're all outlined here, step-by-step, to save you paying a plumber's fee.

The ultimate, of course, is to plan for and build a new 'dream bathroom'. Bathrooms *mean plumbing* but if you decide to build a new bath addition, you should also have at least medium-level carpentry skills. General information is given on framing and the roughing-in of new pipes—enough to give you a good idea of what is involved.

Making Plans for New Baths

Because the bathroom has the most plumbing fixtures of any room in your home, planning for a new bath can be a long and complicated process. Moreover, toilets, the primary fixtures in any bath, present special problems in drainage and venting; so with a new toilet installation, the process becomes even more complex. If you're planning on simply replacing all your fixtures, the job won't be too terrible; but if you're planning on rearranging fixtures, count on some headaches. In most cases it is easier to plan for a whole new room than it is to remodel an old one if you intend to move fixtures around.

By law, you are required to have two checks of your new plumbing by local authorities: approval of the plans when you apply for the building permit and inspection after the work is completed—before the walls are closed up. An exception to this is the three inspections required for a house addition on a slab foundation. In this case, there is an additional inspection which must occur before the floor is poured. You can formulate your own plans and then have them approved by a professional; or you can seek the aid of a builder or architect in the planning process. If you're buying several fixtures from one supply company, you might be offered design assistance for a nominal charge.

In making plans for a new bath, the first thing you should do is read over Chapter 6 so that you have some general knowledge about extending your plumbing. Then, find your existing lines using the guide on page 76. Check the capacity of your water heater before assuming that a large hot-water fixture can be added without depleting the house supply. You might need to add a larger water heater to your system.

Of utmost importance in bathroom planning is the location and the size of the existing soil stack. You can usually find this large pipe in the wall behind the toilet; sometimes it is run in a closet or under a stairway. At the base of the soil stack will be a cleanout plug. Also check for the house drain—the large horizontal pipe that travels from the bottom of the stack to the sewer. New toilets will usually drain directly into the soil stack or have their own secondary soil stack that connects to the house drain. If the crawlspace is adequate, a toilet can have a 3-inch branch

to the main and a 2-inch vent of its own. Generally, toilets must be within 5 or 6 feet of a soil stack or have their own branch drain and vent. Local codes will determine the exact distance.

Piping Options for a New Bathroom. Obviously, bathroom plumbing, requiring such large pipes, is going to be expensive. Therefore, to minimize your costs and your work, you should plan your fixtures relatively close to each other and also close to other fixtures within your home. An ideal situation is to plan for new fixtures on the opposite side of a wall—back-to-back with existing plumbing for another room. If you have such an 'opening' check with your local code to make sure that the existing soil stack is large enough to permit the addition of more fixtures.

If you can't make a back-to-back installation, the next-best thing is to install the new fixtures on another

nearby wall, making sure that the toilet is within a few feet of the existing soil stack. The remaining bath fixtures can be positioned further away from the soil stack, as long as they are properly vented.

Homes are usually designed with as many fixtures as possible branching from the main soil stack. Consequently, in many cases it will be difficult to find a stack that isn't already 'backed up'. When this is the case, you will either have to add a new branch drain or a new soil stack for the proposed bathroom. A new branch drain can be installed under the floor with the vent rising through the wall. The best place to run a new soil stack is inside a new wall, positioning it so that it ends as close as possible to the house drain.

Altering the House Structure. As mentioned earlier, existing soil stacks are usually concealed in extra-thick walls built especially for this purpose. So if you're designing a new bathroom as an addition to your house, you will have to make structural allowances such as thick walls for the piping. The stack, branch drains, and the new supply lines will be run through the walls and floor as in any new construction.

If, on the other hand, the new bathroom is to be built within your present home, the new pipes, in particular the soil stack and branch drain, may be too large to fit within the existing walls. Walls may need to be thickened to accommodate them. For drainpipes and for new supply pipes, you may not wish to tear out sections of the wall for the piping runs. Boxes, closets or cabinets all work to hide your materials (page 84). In any case, take care not to compromise your home's fire stopping capabilities. Do not make excessively large holes in sills or floors to accommodate plumbing. Also, be sure to plug up any unused holes.

Another option for soil stacks is to run them on the exterior of a house, box them in, and finish them off with matching siding. This should be done, however, only in climates where freezing is unlikely. As always, check your code for recommendations and restrictions.

Mapping Out the Fixtures. Once you have determined the location for your new bath, you can begin to work on a plan for the fixtures. This can be done most accurately by making a scale drawing. Using graph paper, mark either ¼ inch or ½ inch for each foot in the room. As a guide for room size, allow at least 5 x 7 feet for a full bath with a tub. You can make the room as large as you want but remember that spreading out the fixtures will add to the cost.

ALLOWING FOR FIXTURE CLEARANCES

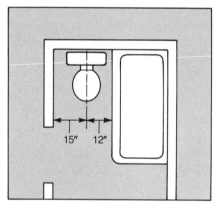

Toilet—Side Clearance. With the center of the toilet bowl identified, the minimum distance permitted to a wall or partition is 15 inches. The minimum distance to a bathtub is 12 inches.

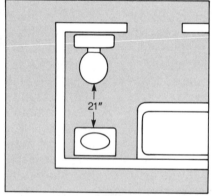

Toilet—Front Clearance. In front of a toilet bowl, the minimum allowable distance to any wall or fixture is 21 inches.

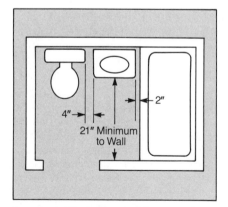

Lavatory Clearance. A lavatory edge can be no less than 4 inches from a toilet tank or finished wall. It can be no less than 2 inches from a tub and 21 inches from its front edge to any fixture or wall.

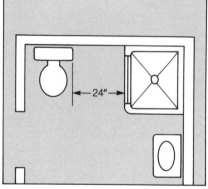

Shower Clearance. A shower stall is permitted to be no less than 24 inches from any fixture or wall.

Map out the room first, complete with windows and doors. Pipes cannot, of course, be run through these areas. Next, map out the location of the soil stack. The position of all the other pipes will be determined by where it is located, where the fixtures are located, and the structure of the existing walls. Check to see where pipes can be run. Fairly easy runs are made with pipes hung under first-floor joists in the basement or crawl space. Another option is to run supply pipes on top of ceiling joists in the attic. In general, try to plan the installation so that pipes run between floors will lie parallel to and between joists.

After shopping for fixtures, gather the dimensions of each fixture that you intend to put in the bathroom. Cut out small scale models of the fixtures and then place them on your scaled plan. Move them around to the desired positions, keeping in mind that there are code regulations about placement. Fixtures must be properly spaced from each other to allow for access when they are repaired or cleaned. Though local codes vary slightly, typical clearances for fixtures are shown here (left).

Planning the Details. As mentioned on page 77, all soil pipes must be sloped downward toward the stack at ¼ inch per running foot of pipe. In addition to this rule, never plan to install a major fixture, such as a toilet, upstream from a minor one. A lavatory, bathtub, or shower must have a trap whereas a toilet will have a closet bend that slopes directly toward the stack. Fixtures must be vented. (See page 76 for the different types of venting.) The National Plumbing Code recommends that a tub or shower have a trap within 3½ feet of the soil stack. They also advise that a lavatory within 2½ feet of the stack be wet vented, but check your local code for variations.

Plan for the supply pipe runs using the information in Chapter 5. Use fixture templates and plan to run the pipes at least 6 inches apart from each other. Though supply pipes are not required to be sloped, doing this will be beneficial when you want to drain the pipes; slope them approximately the same as the DWV pipes. Supply pipe should always end at the fixture with the cold water to the right and the hot on the left.

Choosing Materials. For your DWV system, use ABS pipe if your code permits it. It is the easiest to work with and even if your existing pipe is cast iron or copper, this plastic pipe can be connected to it easily. The required size as determined by code is calculated by fixture units (page 77). Remember also that the drainpipe's length is also regu-

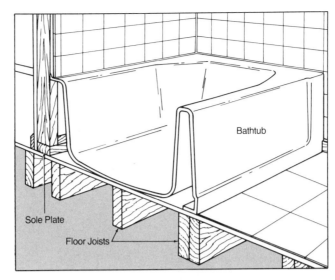

Framing in a Bathroom.
Floor joists should be positioned with the plumbing and wiring in mind. Single joists running parallel to the wall are separated to allow for pipe and wire runs. Double joists on the outside edge of the bathtub provide extra support for this heavy fixture.

(Labels in illustration: Bathtub; Sole Plate; Floor Joists)

lated (see the 'Critical Distance', page 80). Soil stacks are generally 3 to 4 inches in diameter and branch drains are 1½ or 2 inches. However, the trend is toward smaller pipes, so always check your local code first; then use the smallest pipes that the code permits.

Copper tubing or plastic pipe should be used for the supply lines. Copper connects to all kinds of existing plumbing and is accepted by all codes. Plastic is not accepted in all areas.

Bathroom Installation Tips

Once you have finally made your plans and had them approved by local building authorities, you can begin your work. Be sure to read over Chapter 5 thoroughly and also review the rest of this chapter which details individual bathroom fixture replacements. If you are replacing fixtures, there should be no problem doing all the work yourself. If you are constructing a new bathroom, however, you should either have medium-level construction skills or seek the help of an experienced carpenter.

General Construction Tips. If you are building a new addition for a bath, doing the carpentry and plumbing in a logical sequence can save you a lot of unnecessary labor. Once the posts and girders are in place and the foundation has been poured, you can begin the plumbing. Often plumbers will wait longer, however, until after the floor joists are in.

Proper framing in a bathroom can help immensely when the time comes to rough-in the piping, and it can also make for a smoother installation of your electrical wiring. Most importantly, floor joists that run parallel to walls should not be set directly under them (see the illustration, above). Instead, they should be placed on either side of the

wall. That way, you will avoid cutting through them when you install drainpipes for the toilet, tub, and shower. Additionally, to provide extra support for a bathtub, you may frame with double floor joists on the outside edge of the tub—away from the wall.

Roughing-In DWV Pipe. Follow these basic steps when you are installing an entirely new bathroom. First, install only the first 5 or 6 feet of the soil stack and cleanout. If it and the house drain are different sizes, use an adapter or reducer to connect them. Wait until after the framing is complete to install the rest of the soil stack.

Next, locate the exact position for the toilet drain hole and cut out a hole slightly larger than the fitting that will go here. The floor flange under the toilet (page 51) connects to a closet bend that turns toward the soil stack. Make sure that this pipe is well supported. If it isn't, add framing members for stack clamps. Add the additional drainpipe, working backward from the soil stack to the fixture. Support horizontal pipe as directed by your code.

The next step is to put in drains for the tub, shower, and lavatory. Install Ts as needed for drainpipe and venting. Stub out all drains and wait until the walls are installed before adding fixture traps.

Roughing-In Supply Pipe. As shown on page 79, use fixture templates for determining fixture stubouts. Run the pipes at least 6 inches apart, working with the main lines first and then the branch runs. To minimize pipe movement from temperature changes or hammering, support the pipes at 4- to 10-feet intervals. If the code requires it, install air chambers above fixtures where branch runs tie in with a T fitting.

REPLACING A TOILET

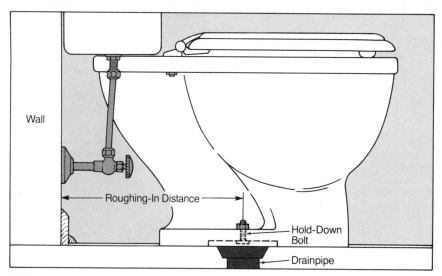

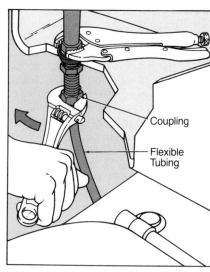

Wall

Roughing-In Distance

Hold-Down Bolt

Drainpipe

Coupling

Flexible Tubing

1 Determine the roughing-in size—the distance from the wall to the center of the drainpipe. Do this with the old bowl still in place, by locating the hold-down bolts that secure the fixture to the floor. Measure from the center of these bolts straight back to the wall behind the bowl; this gives you the exact location of the hidden floor drainpipe. If the toilet has two pairs of hold-down bolts, measure from the rear pair. The rough-in distance of the new bowl can be shorter than this dimension, leaving you with a gap behind the wall, but it cannot be longer; otherwise the bowl would not fit.

2 At the bottom of the old tank, unfasten the coupling on the flexible tubing. If it is kinked or corroded, replace it with a new piece.

CAUTION

Before installing a new toilet, turn off the water at the fixture shutoff valves or the main shutoff valve (page 9). Then flush the toilet twice to empty the tank and bowl. Remove any remaining water with a sponge or rags.

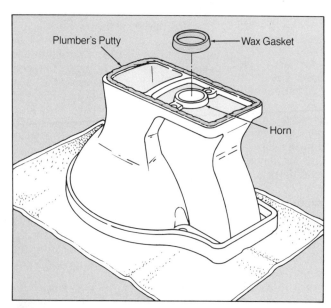

Plumber's Putty

Wax Gasket

Horn

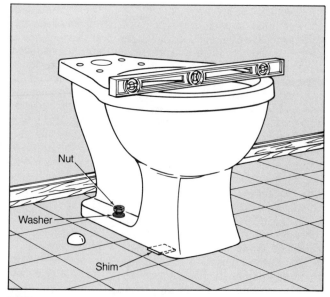

Nut

Washer

Shim

5 Carefully turn the new bowl upside down onto a cushioned surface. Position a wax gasket around the horn (water outlet) of the bowl. Apply plumber's putty around the edge of the bowl's bottom.

6 Take the rags out of the drainpipe and carefully lower the bowl into place over the flange, using the bolts as guides. Press down with a slight twisting motion to seal the bowl. Place a level on top of the bowl and check for tilting. Use thin metal shims to level the bowl if necessary. Put on new washers and nuts at the hold-down bolts; hand-tighten only.

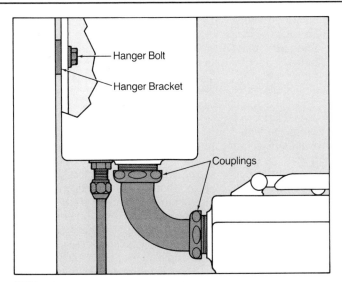

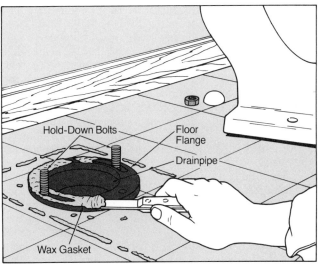

3 For a bowl-mounted tank, detach by unfastening the mounting bolts (page 56, step 1).

For a wall-mounted tank, loosen the couplings that connect the pipe to the bowl and tank (page 56, step 1); use a spud wrench. Have an assistant hold the tank in place while you unscrew the hanger bolts and remove the tank from the wall. If the new toilet will not be wall-mounted, remove the hanger brackets.

4 Pull off or unscrew the porcelain caps from the floor bolts. Using an adjustable wrench, remove the nuts and washers. If they are corroded, soak them in penetrating oil. In order to break the seal between the floor and the bowl, gently rock the bowl back and forth. Once broken, lift it straight up and set it aside. To keep unpleasant sewer gas from escaping, and to prevent objects from falling in, stuff an old rag into the drainage hole.

With a putty knife, scrape up the old wax gasket. Remove the old hold-down bolts from the floor flange and thoroughly clean the flange for use with the new bowl. If it is broken or cracked, replace it with a copper or plastic model that can be soldered or cemented into place. Make sure that the bolts are positioned correctly.

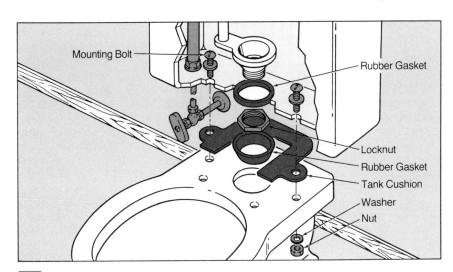

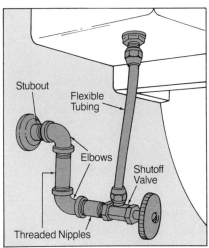

7 To mount the tank, first fit the rubber gasket onto the flush-valve opening on the bottom of the tank. Next, put the rubber tank cushion on the rear of the bowl. Position the tank onto the bowl and tighten the locknut and nuts and washers onto the mounting bolts. With wrench, snug up the hold-down nuts at the base. Check to see if the bowl is level. Fill the porcelain caps with plumber's putty and position them over the bolts. With caulking compound, seal the base of the toilet to the floor.

8 To connect the water supply it might be necessary to readjust the fittings for the supply pipe or install new fittings. To bring old fittings to the desired position, screw an elbow onto the wall stubout. Use two 4- to 6-inch threaded nipples and a second elbow to connect a shutoff valve. Install the flexible supply; the flared end will be connected to the ball-cock (page 55). Tighten all connections, turn on the water and check for leaks.

Tips on Toilet Installation.
Toilets require their own vent going through the roof (usually 2 inches) and a drain that is at least 3 inches in diameter. If a toilet is on a branch drain, it can't be upstream from a shower or sink. Rough-in the closet bend and the toilet floor flange first. Position the flange to the level of the finished floor. The pipes required include: closet bend and drainpipe, cold water supply stubout with shutoff valve, flexible tubing above shutoff valve and, possibly, an air chamber.

Tips on Lavatory and Bidet Installation. Lavatories may be installed back-to-back, side-by-side, or within a vanity cabinet which will allow you to expose the pipes. If a sink is within the critical distance, it can be wet-vented; otherwise it should be back-vented. Required pipes include: drainpipe and drain stubout, hot and cold supply stubouts, shutoff valves, and flexible tubing above shutoff valves. Optional materials are transition fittings and air chambers.

Bidets, fixtures for washing the perineal area of the body, are installed in the floor. However, they don't require the same plumbing as a toilet; no closet bend is necessary. Instead, they have a sink-type trap and drainpipe.

Tips on Bathtub and Shower Installations. Bathtubs and shower stalls are often positioned on branch drains and they may be back-vented or wet-vented. Because of the floor drain trap, both types enter the stack at the floor or below. Support framing is usually required (see page 97 for bathtubs and 104 for showers). While the wall is open, install the shower faucet body and shower head. Pipes needed for a combination tub/shower installation include: drainpipes and traps, hot and cold supply pipes, and a pipe to a shower head. An air chamber may also be required.

Mounting a New Toilet

Installing the piping for a new toilet can be tricky but removing an outdated or faulty toilet and setting in a new model can be relatively easy—a project taking you no more than an afternoon to complete. The new bowl simply fits over the existing floor flange and drainpipe. The existing supply pipe can also be hooked up to your new tank. This can be done even if the old model is an 'antique' wall-mounted model.

When you shop for a new toilet, be sure that you have measured accurately for the 'roughing-in distance'—the distance from the wall to the center of the drainpipe (page 98). Your new toilet must be small enough to fit in this space. Other than the correct size, additional buying considerations are water conservation and cost. Newly designed models with lower tanks and narrower bowl traps use about a third less water per flush than conventional toilets, but the installation is no different.

All of the internal mechanisms of the tank will probably be supplied with the fixture, as well as the hardware for fitting the tank to the bowl. The only things that you might have to purchase separately are the hold-down bolts, the new wax gasket or a can of bowl-setting compound for sealing the bowl to the floor.

If you need to reroute your water supply pipe for a connection to the tank, use flexible connecting pipe. Be sure to specify that you need toilet-tank supply pipe—a different type than what is used at sinks or lavatories. Also, purchase and install a shutoff valve if your old toilet did not have this convenient fixture (pages 84-85).

You'll need some extra strength and possibly assistance when removing the old and mounting the new toilet. Be sure to do the mounting step very carefully as porcelain chips easily when it's dropped. The tools needed for the job, however, are simple and few: a large pipe or monkey wrench or a spud wrench, a tape measure, screwdriver, level, and putty knife.

Bathroom Lavatories

There are several types of lavatories as determined by their structure. Deck-mount types are shown on page 89 and the instructions for installing them are also given (page 88). There are two other basic types that will be covered here: floor-mount and wall-mount lavatories.

Old-fashioned floor-mount models are sealed to the floor with Plaster of Paris while wall-mount lavatories are hung on a wall with hanging devices. Basic removal of both kinds is no different, however, than that for other sinks. With the water turned off and drained, disconnect the old faucets and the drain. Follow the steps at right to finish the job.

Tips for Installing New Lavatories. Use the instructions on pages 101-102 to remove an old and install a new wall-mount sink. It's recommended, however, that you consider replacing your old sink with a deck-mount type. These are safer than wall-mount models and generally more esthetically pleasing. If you're extending pipes for a new sink, use the instructions on pages 75-83 to make the run. You can use the same basic instructions as well as the general 'bathroom' information on pages 95-97 to install a new sink alongside your old one.

When purchasing your new lavatory, also select a new 4-inch or 8-inch faucet to go with it; the option is to salvage an old one. In most cases, the faucet should be installed in the lavatory before it is set into place. If you're putting in a wall-mount lavatory, it can be positioned at any comfortable height, although the maximum recommended height is 38 inches. The standard height for a deck-mount model is 31½ inches.

Replacing an Old Bathtub

This process will be quite involved because it not only requires muscle to haul out the old tub but it also calls for fairly advanced plumbing skills to make the proper connections. So you should read over this and all of the information in this book regarding tubs (pages 40 and 47) and showers (page 49) so that you'll understand what you're up against. This might be a project that warrants hiring a plumber. As an option, you could do the dismantling yourself and hire a professional to do the installation work.

If you're making a replacement, measure the old tub accurately and shop for a new model of the same size. Tubs are made of four different materials—fiberglass, acrylic, porcelain-enameled steel, and enameled cast iron—and they all have their disadvantages and benefits.

The most expensive and most durable is cast iron. The porcelain-coated steel tub is lighter and easier to install but it tends to chip easier. Less expensive and easier to install are the lightweight fiberglass and acrylic types.

Maintenance-wise, however, they fall short. In time, they will show scratches and loose their glossy surfaces. Even worse, if you or someone else mistakenly uses a gritty cleanser on them, the surfaces will be permanently damaged.

When shopping for your new tub, take the dimensions of the area and also a complete detailed plan of the existing pipes and the surrounding supporting joists and other framing. Have a representative of the plumbing supply store or a designer help you select a model that will require the least amount of change in rebuilding your bathroom walls.

Removal and Installation Tips. Free-standing tubs, the old models with legs, can be simply picked up and hauled out of the bathroom. Obviously you'll need assistance, and plenty of it, to perform this feat. Old cast iron tubs can be broken up while steel models need to be carried away in one piece. To tell the difference between the two, tap it with your knuckles. A cast iron tub will sound solid like a rock whereas a steel tub will 'ring' or sound hollow. During the removal process, wear goggles and protective clothing to shield yourself from flying chips and flying pieces.

One important point to remember is the size of the tub; always measure your house doorways and stairways to make sure that both the old and the new fixture can pass through them. If your new tub is smaller than the space created

by the removal of the old one, you can build out the wall to meet it. Such an inset can also be used as a shelf for shampoos and other bath products.

Very old tubs may have the faucets set into the tub; if this is the case, then you'll have additional plumbing to do. New tubs come with only two holes: one for the drain and the other for the overflow. Piping for the spout, faucets, and shower will be above the tub; it should all be installed before the tub is set into place. Tubs are usually placed on the subfloor while the house is being built and then the flooring is laid so that it contacts the fixture.

Like all plumbing fixtures, your new tub will come with a template and detailed instructions for the installation. Read and understand this information thoroughly before you even begin. Use these plus the general steps for a typical installation on page 103. The elements include: the hot and cold water valve assembly, hot and cold

supply piping, the shower pipe, pipe stubs for the tub spout and shower head, and a diverter valve. The latter turns on the shower and is located either between the faucets or on the spout.

As always, check with your local building code about the installation.

CAUTION

■ Older sinks can be very heavy. The wood holding hanging devices could be rotten and the sink might be supported by plumbing pipes. For all of these reasons it's best to work with a helper who will support the sink while you disassemble it.
■ Before removing a lavatory, turn off the water at the fixture shutoff valves or the main shutoff valve (page 9). Then disconnect the supply pipes.

REMOVING BATHROOM LAVATORIES

Removing a Floor-Mount Lavatory.
Loosen any bolts that hold the basin to the top of the pedestal and lift off the basin. Then rock the pedestal back and forth to break the plaster of Paris holding it to the floor. For a sink that is resistant, wrap the base with rags to protect from flying chips and pound with a hammer at the base. Sand or scour the floor to remove the plaster residue.

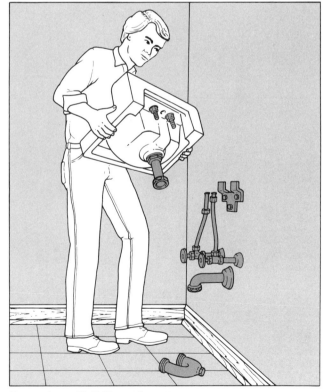

Removing a Wall-Mount Lavatory.
Disconnect the trap from the tailpiece (page 46). Underneath the sink, disconnect the supply pipe couplings at the faucet inlet shanks. Examine your sink for any hold-down bolts. If necessary, unfasten. Pull straight up to lift the sink off the hanging device. If the faucet and drain assemblies are in good condition, remove them.

Some codes require an access to the drainpipe and supply pipes. If your tub is on the ground floor, this can be from your basement or crawlspace. A hole measuring approximately 1 foot by 8 inches should be adequate; cut it out of the subfloor at the head of the tub. An option (or a necessity if the tub is upstairs) is to build an access door or removable panel into an adjacent wall. These can be located in hallways, closets, or adjoining rooms.

Shower Installations

Common shower stall problems include cracked tile, loose grout, or a leaking shower pan. The grout and tile problems can be easily fixed and obviously require no plumbing. A faulty shower pan, however, is quite another problem. Chances are that the leak has been around for some time and it's likely to have done damage to the framing structure of the house. To be sure about what repair route to take, you need to remove the shower pan, inspect the supports and repair any parts that need it. If the dismantling is extensive, you should probably replace the entire stall.

If your existing shower stall has a tile floor or you want to install a tile floor in a new stall, be sure to check with your local code first. In the not-too-distant past, codes were particularly stringent regarding this installation—the reason being that tile floors were always, in time, subject to damaging leaks. Products such as lead, copper, or hot-mopped shower pans were required.

INSTALLING A NEW WALL-MOUNT LAVATORY

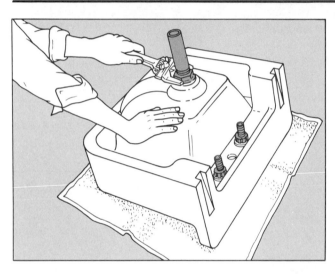

1 As shown here and using the directions on pages 46 and 89, prepare the sink by attaching the pop-up stopper, the faucet, and the tailpiece.

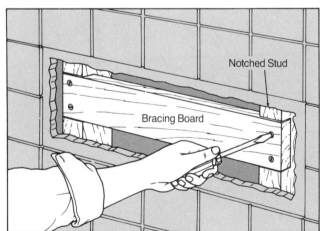

Notched Stud

Bracing Board

2 For a first-time installation and as a security measure, install a bracing board into the wall. If your wall has ceramic tile on it, remove just enough tile for a 1 x 4 board to be inserted from one stud to another. Cut notches into the studs deep enough for the board to be flush to the wall surface, minus any drywall or tile finish. Screw the backing board to the wall and then refinish the wall with drywall or tile. Attach the hanging device to the wall using the instructions above plus any manufacturer's directions.

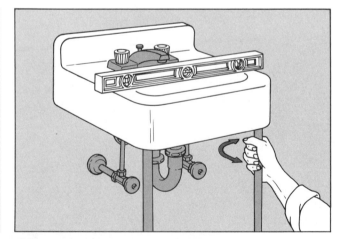

3 Carefully lower and place the lavatory on its hanger. If the sink needs additional support, brace the front of it with a 2 x 4. Then attach the supporting legs. If the legs will rest on tile, you can use a carbide bit to drill ¼-inch-deep holes into the tile. Fasten the legs to the front of the sink and screw the adjusting sections downward until the sink is level. Remove the supporting 2 x 4.

Seal the sink-wall joint with caulking compound. Connect the water supply pipes and the trap. Turn the water back on, check for leaks and tighten any leaky connections.

INSTALLING A NEW BATHTUB

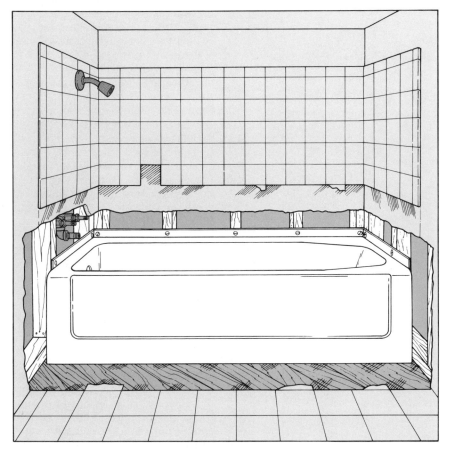

1 Begin by measuring the tub and the doorway and planning for the exit. It might be necessary to remove the door or the door trim. Do any such dismantling first. Start the tub disassembly by removing the overflow plate, the pop-up or strainer (pages 49-50), and the spout (page 48). Then unfasten the supply pipes and the drainpipes. Chip away any tile or drywall that meets with the edges of the tub leaving nothing but studs visible several inches above it, as shown. Also, remove any flooring that meets with the tub; strip it down to the subfloor.

If you have a cast iron tub, you can break it up with a sledge hammer and carry it away in pieces. If you wish to leave it in one piece, have at least four people with you to assist in handling it. Steel tubs cannot be broken up; they usually have nails or screws at the top of the flange to hold them into place. Remove this hardware; then, lift out and carry the tub away.

CAUTION

Before removing a bathtub, turn off the water at the fixture shutoff valves or the main shutoff valve (page 9). Then open the faucet to drain the pipes. While dismantling, be sure to wear gloves, heavy clothing, steel-toed shoes, and eye protection.

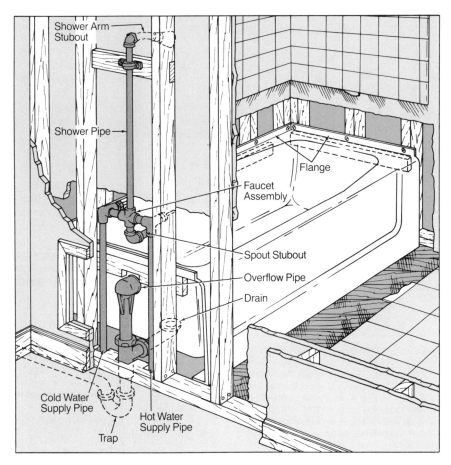

Shower Arm Stubout

Shower Pipe

Flange

Faucet Assembly

Spout Stubout

Overflow Pipe

Drain

Cold Water Supply Pipe

Hot Water Supply Pipe

Trap

2 Clean the area of all debris and clean off the exposed studs. There should be two precut holes in this area—one for the overflow and one for the faucet. If the new tub is cast iron, it will simply sit in the enclosure. If the tub is made of steel, install the needed 1 x 4 or 2 x 4 supports. Use nails or screws to anchor the tub through the flanges into the studs. Do the piping for the drain, overflow, faucet, spout, and shower head. Make the drain assembly connections through an access door or through an opening from below. Once everything is connected, turn the water back on and check for leaks. Then finish the wall using moisture-resistant drywall as a base. After the final wallcovering is applied, seal all joints with silicone caulk; you might consider a mildew-resistant substance if moisture has historically been a problem in your bath. Lastly, install the spout, faucet handles, and the shower head.

PUTTING IN A NEW SHOWER STALL

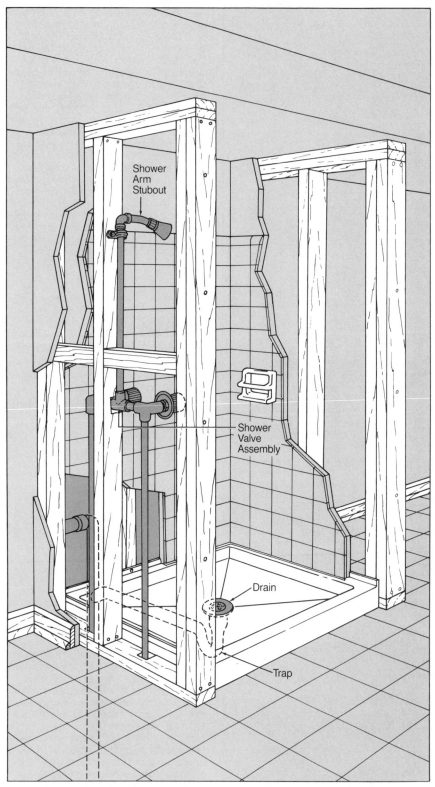

Construct a frame for the shower unit being careful to measure, cut, and assemble precisely. Make sure that the unit is square and plumb. Run drainpipes and supply pipes to the location and install pipes for the faucet, spout, shower head, and drain. Position the stall into place following manufacturer's instructions. If necessary, nail the flange to studs, a step similar to that used for the bathtub installation (page 103). Again following specific manufacturer's instructions, drill the holes for the shower valve assembly and the shower arm stubout. On some models you must drill into the surface of the stall; on others the holes are predrilled. Install the shower head, faucets, and drain using the information on pages 49, 47, 40, and 46, respectively.

Today's materials have changed all of that and now sturdy molded plastic bases are the norm. Besides being 'leak-proof', these are also versatile; they can be used with metal and fiberglass shower stalls as well as tiled models.

When you shop for a new shower stall there are several things to consider. Among them are the location of the new fixture, the maintenance required, and the cost. If you're installing it in a remote area of your home, you will probably go for a less expensive metal type. But if you're making your installation adjacent to a bedroom, as part of a home spa or in some other showplace, you'll probably opt for the more expensive tiled-wall or fiberglass model.

Whichever type you choose, it will come with explicit installation instructions. Use the information given here, combined with the manufacturer's instructions, to make your plan. When doing the plumbing, be sure to place the piping accurately. Use flexible copper tubing for the supply lines; this way, minor adjustments can be made as you set the new stall into place.

If you have chosen a tile stall, select a plastic shower base of the same size and then frame the stall to fit it precisely. Connect the base to the drain line and then cover the framing with water-resistant (usually green) drywall. This should cover the lip at the top of the base no more than ¼ inch away from the curb. Extend the drywall up at least 6 feet high and then cover it with tile. Setting in the tile is a relatively easy process; just be sure to use a waterproof mastic. An alternative to the green drywall is fiberglass reinforced light-weight concrete board to which tile is adhered by the 'thin-set' method.

Hand-Held Showers

Hand-held showers have several advantages. They can be mounted in the usual position to spray water as a normal shower does or they can be used by the bather to direct water wherever it's needed. In this way they're convenient and versatile but they also offer another benefit. Because most of them are made with plastic instead of metal parts, sediment buildup is minimized.

Since these fixtures are relatively inexpensive, to make your maintenance easy, you should purchase a high quality brand if you can afford it. Get one that disassembles easily, has available replacement parts, and has concise instructions for repair.

Three basic types are available. The first (not shown) simply replaces a standard shower head. It has a built-in holder for when you want to leave it stationary; to remove it for hand-held use, you simply lift it out. Installation for this model is much like the procedure on page 49.

The second type allows you a choice of using the old showerhead or the new hand-held unit. For this installation you must first attach a new diverter valve at the shower arm and then attach the unit to it.

The third type of hand-held shower can actually add a shower where there previously was none. When one of these is installed, the original tub spout must be replaced with a special tub-spout diverter to prepare for the hookup. Detailed instructions are given below for installing the last two types of hand-held showers mentioned.

CAUTION

Before installing a new hand-held shower, turn off the water at the fixture shutoff valves or the main shutoff valve (page 9). Then open the faucet to drain the pipes.

ATTACHING A HAND-HELD SHOWER

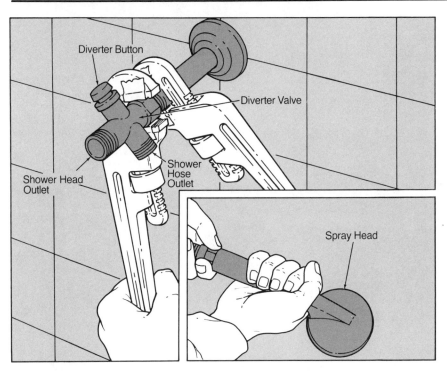

Diverter Button

Diverter Valve

Shower Head Outlet

Shower Hose Outlet

Spray Head

To a Shower Head. Begin by removing the old shower head using tape-wrapped pliers and the directions on page 49. Clean the threads on the shower arm with a wire brush; apply pipe-joint compound or mylar tape. With two wrenches, thread your new diverter valve onto the shower arm. Tighten the valve until the diverter button faces upward. Fasten the hose for the hand-held shower to the shower head outlet and fasten the old shower head to the shower hose outlet.

Screw the hose into the hand-held spray head. Install the wall bracket for hanging the assembly. Turn the water back on and check for leaks.

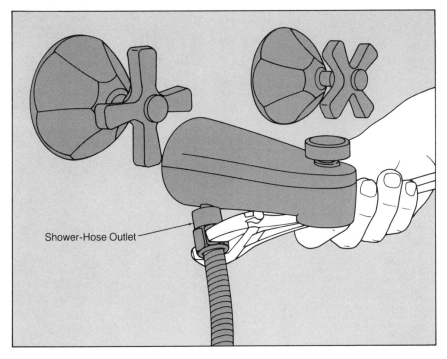

Shower-Hose Outlet

To a Tub Spout. For this connection you'll first have to replace the tub spout with one that provides an outlet for the new shower hose. Begin by removing the tub spout (page 48) and replacing it with your new spout. Apply pipe-joint compound to the threads of the shower-hose outlet and install the hose. Attach the spray head to the hose as shown above. Install the wall bracket for hanging the assembly. Turn the water back on and check for leaks.

Cimarron

Energy Efficient

Five Year Warranty

COMPLIES WITH ASHRAE STD. 90A-1980

MANUFACTURED FOR SUTHERLAND LUMBER CO. (K.C. MO)   BY MOR MFG. CO.

Plumbing in Utility Rooms and Basements

Utilitarian areas of the home are in different places depending on your home's design. You might have an old unfinished basement, a small centrally located space for your water heater, or anything in between. The one thing all these areas have in common, however, is that they're usually 'out of sight' and, consequently, pipes are exposed. This ensures that plumbing will be easier to install here than elsewhere in the home where pipes must be concealed.

Two appliances often located in the utility room or the basement are the clothes washer and the water heater. Neither of these 'monsters' go on the blink often but when they do, a domestic crisis can occur. A breakdown in your washer can be quite exasperating as laundry piles will escalate all over the house. Routine maintenance can help to keep washer problems to a minimum

and when it's time to put in a new model, you can do the plumbing.

Troubles with a water heater are a little more subtle—you're likely to have a warning, such as a leak, which you should take heed of. Step-by-step instructions are given for installing a new heater, plus tips are given on regular upkeep. Just remember, these appliances combine plumbing and electricity, or, in the case of the water heater—gas, so extra caution must be used when working around them.

Another fixture that's often found in the basement or utility space is the water filter. Installing one here is likely to be a little easier than putting it at your kitchen sink, where workspace is limited. Other topics discussed are putting in a sump pump and, also, basement toilets.

Major Cleaning Machines— Clothes Washers

Automatic clothes washers are major home appliances; that is, they are used a lot, especially by large families. Consequently, they are sturdy, well-built machines. Clothes washers are composed, among other things, of electrically operated solenoid valves and high-torque electric motors that start and stop frequently as cycles change. Because of

this, a washer requires a 15-amp grounded circuit, preferably with a delay-type fuse or circuit breaker at the service entrance panel or fuse box. This kind of device will cut down on blown fuses or tripped circuits when the motor causes frequent power surges.

Hot and cold water lines are required for washers and these supply pipes should measure at least ½ inch. Washer drainpipes are required to be 1½ inches

in diameter. Many homes have laundry tubs into which a washer's wastewater is drained and, if that is the case, the laundry tub must likewise have a 1½-inch drainpipe.

If you have no laundry tub available, however, you will instead have to use a *standpipe*. This tall 2-inch-diameter pipe is needed because of the speed with which water is pumped out of the washer. If a standpipe is connected to a floor drain, that drain must have a trap in it. If it is connected to a vertical drainpipe, as shown below, the trap will be built into the standpipe. A trap must be incorporated somewhere into the washer drainage system. There are additional code regulations about the drainpipe into which a standpipe is installed; toilets or tubs cannot be located further up the line and no fixture other than a toilet can be installed further down the line.

When clothes washers are located on the first floor of a house, standpipes are sometimes hidden inside a wall. A commonly used device in such an arrangement is a metal panel or inset. Box-like, this has an open front that provides for draining away any leakage from the supply connections as well as any small amount of water that might back up from the drain line.

Because the washer's solenoid valves open and close so quickly, water hammer is likely to occur in the washer's supply lines. While it is true that the washer's rubber hoses will absorb a great deal of the water hammer, they will not absorb all of it and, furthermore, constant hammering will wear out these hoses. For this reason, you should install air chambers or shock absorbers on both the hot and the cold supply pipes.

Whenever you suspect any electrical problem with your clothes washer, it's best to call a repairman. There are, however, a few things that you can do to keep your washer operating properly. Hose connections should be checked periodically for leaks and the hoses themselves should be checked for cracks. If you have any trouble with the machine not filling properly, you can check the supply connections for sediment that might be blocking the flow. Remove the hoses at the valves and inspect the screens; clean them or replace them if they look bent or worn.

The best maintenance procedure that you can take for your washer is to always turn off the supply valves when the machine is not in use. If you don't take this measure, your hoses will be under constant pressure and so will the electric solenoid valves. If the shutoff valves for your washer are located behind it, this inconvenience might be cause for changing the piping.

Putting in a Standpipe. Standpipes should be installed so that they are taller than the highest level of water in the washer. This will prevent a backup and siphoning of dirty water into the machine. The pipe must be a minimum of 1½ inches in diameter but

you should check the washer manufacturer's recommendations for the specific dimension. Most washers are made with a bend in the drainpipe that ends in a 6-inch hook; this feature will ensure that the pipe will stay in the standpipe (or the laundry tub) and not be worked out by the force of draining water.

Preparing Pipes for a New Washer. Typically, building codes require that new residential units have washer connecting pipes already installed. However, if you live in a home that has never had a clothes washer or, if you wish to locate your washer elsewhere in your house, you should be able to do the plumbing without major trouble. After you have made your plan for new piping runs, however, be sure to check with your local building department for approval.

Instructions are given (below, left) for a typical piping installation complete with air chambers above the valves. As for the valves, you should use separate hot and cold valves, the threaded-spigot types, for attachment of the washer supply hoses.

The Home Water Heater— Workhorse of the Hot System

Water heaters are relatively simple plumbing devices but that doesn't mean that you should be careless when working around them. If they are not properly equipped and maintained, these large tanks can be the most dangerous plumbing fixture in your home. Because most tanks are gas- or electric-powered, you should use extra caution when working on them.

The basic construction of a hot water heater involves a vertical tank that is usually 'glass-lined'. The lining is actually a type of porcelain that is fused to the steel inside the tank to prevent rust and corrosion. The three basic types of heaters are: gas-powered, electric-powered, and oil-fired. If a water heater is gas-powered, it will have a flue inside it through which combustion gases are released via a house chimney. Most of the heating is done at the bottom of the tank where a large burner, much like that on a gas range, is located.

Electric water heaters have no flue. Instead, they have two electric heating elements inserted in the wall of the tank

INSTALLING A STANDPIPE

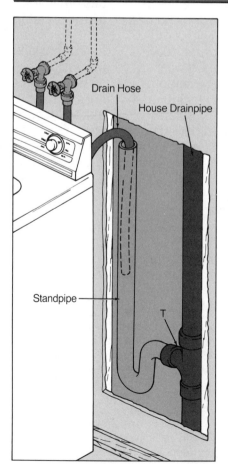

Drain Hose

House Drainpipe

Standpipe

T

Standpipes with traps are available in lengths of 34 to 72 inches and with diameters of 1½ to 2 inches. Check the washer manufacturer's recommendations for what size to buy. A standpipe can be installed into a floor drain or connected to a vertical house drainpipe as shown here. In any case, don't ever hook up a drain hose directly to a house drain; an air gap must be provided to eliminate the possibility of back-siphonage into the machine. For this installation, the house drain may not be connected to a tub or toilet above or to any fixture other than a toilet below it. Use the directions on page 81 for installing the T into the house drainpipe. Insert the drainhose from the washer into the standpipe far enough that the water pressure from the emptying washer will not cause it to work its way out.

by threaded watertight connections. Oil-fired tanks have a blower that is a smaller version of the type used in oil-fired furnaces. Heat is supplied at the bottom of the tanks as well as on its sides and a flue is required for the gases that are emitted during combustion. Water heaters fueled by oil are not very common these days; they are usually seen in homes where oil-fired furnaces are used.

One of the important features of all water heaters is their *rate of recovery*—the time in which it takes to heat water to a preset temperature once an amount is drawn from the tank. Gas and oil heaters have fast rates of recovery whereas electric tanks are slower to reheat. Because of this, electric tanks will usually need to be larger than the other types—when serving the same size homes. Tanks are available with 30- to 82-gallon capacities and the size should be determined by the number of bedrooms in the house. Thirty-gallon (gas) tanks are normally used in one- or two-bedroom homes; 40-gallon (gas) tanks are used in three-bedroom homes, and so forth. Slightly larger electric tanks will be used in the same homes.

Water Heating Alternatives. The fact that energy conservation has become such an important issue in recent years has prompted many people to search for alternative methods in heating their home water supplies. One device that is coming back into fashion is the *demand* water heater. Also called *instantaneous* or *tankless* water heaters, these are very common in Europe and Japan. Because they work on demand, they use up to 20 percent less fuel than the usual tank types.

Another energy alternative that has been gaining in popularity is the use of solar water heaters. Depending on your climate, these can be great investments, paying for themselves in less than ten years in most cases.

The best time to plan for solar heating is when you're building a new home. In general it's best to study the subject very well and have a professional do your design work.

Hooking Up a New Water Heater

You'll know that the time has come to replace your current water heater when it starts leaking or when you see evidence of rust and corrosion. As you shop for a new one, there are several things to keep in mind.

The warranty supplied by the manufacturer is of utmost importance. Heaters come with 7- to 15-year warranties and it's wise to select the best that you can afford. Also, make sure that you get a heater that has the correct capacity for your household; if you intend to expand your family or install new fixtures in the future, this is one way that you can prepare. Along the same lines, purchase a model that has a recovery rate suitable for your lifestyle. This is best tested when several loads of laundry have just been done and all the family members take showers in succession.

As far as the fuel supply, unless there is a significant cost difference, it's best to stay with the same type of fuel that supplies your current tank. Instructions are given on page 110 for replacing a gas-powered and an electric-powered water tank. For both of these installations, use caution; *be sure to shut the gas or electrical power off!*

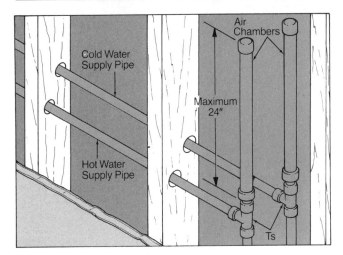

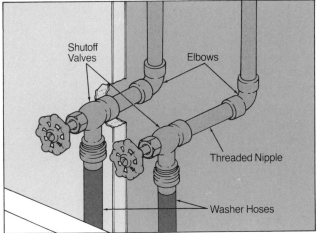

1 Read the manufacturer's recommendations for installation. Most local codes require ½-inch supply pipes. Run supply pipe from the nearest location. Extend the piping to a location just above the washer. To the ends of both pipes, install T fittings. From the lower part of the fittings, extend pipe for the shutoff valves; from the upper part of the Ts, extend pipe for air chambers. A typical recommended size is a diameter one size larger than the diameter of the supply pipes and a length of up to 24 inches. Install reducers, if necessary, and install capped pipes for use as air chambers.

2 Extend pipes from the installed Ts to a position above the washer that will allow for separate shutoff valves. Install two elbows, one for the hot water line and one for the cold. To the elbows, connect threaded nipples. To the nipples attach shutoff valves with threaded spigots; the washer hoses simply screw into the ends.

CAUTION

Before installing the plumbing for a clothes washer, turn off the hot and cold water at the nearest shutoff valves or the main shutoff valve (page 9). Then drain the pipes.

INSTALLING A NEW WATER HEATER

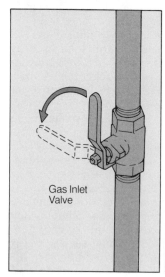

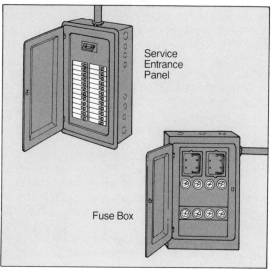

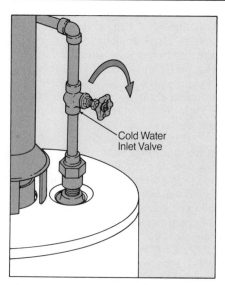

Gas Inlet Valve

Service Entrance Panel

Fuse Box

Cold Water Inlet Valve

1 Begin by shutting down the power supplied to the water heater. If you have a gas heater, locate the gas inlet valve, usually on the side of the tank, and turn it off as shown or as indicated by the valve OFF/ON markings. If you have an electrical heater, switch off the circuit breaker or pull the appropriate fuse at your home's main service entrance panel or fuse box.

2 Next, turn off the cold water that enters the tank. The inlet valve is located at the top of the heater. Open a hot water faucet or a fixture located some distance away from the heater to begin the draining process. When drained, close the faucet. Allow the tank to cool down for a while and then empty it. Let water run into a floor drain underneath the tank or, if none exists, connect a hose to the valve and run it to a nearby drain or outdoors. Open the drain valve near the base of the tank and allow it to empty.

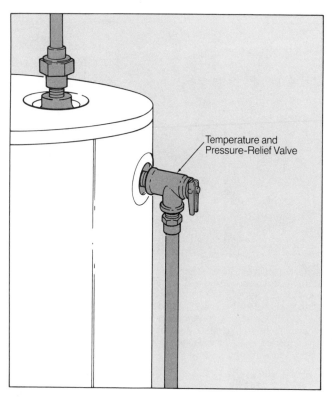

Temperature and Pressure-Relief Valve

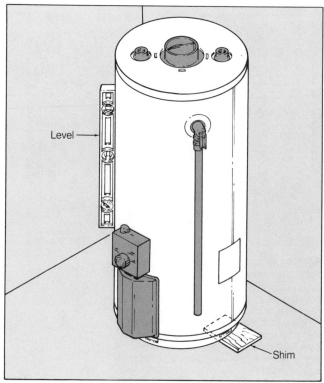

Level

Shim

5 Install the new temperature and pressure-relief valve in the new tank by following the manufacturer's instructions. Make sure that the valve is compatible with the tank; don't ever use an old valve on a new heater.

6 Remove the old heater and move the new one into place. Check it with a level to make sure that it is plumb. To correct a tilted heater, shim it underneath with wood scraps.

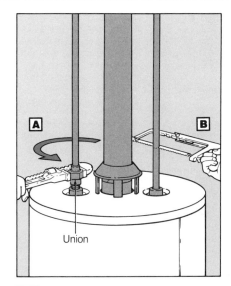

Union

3 Disconnect both of the water pipes—the cold water inlet pipe and the hot water outlet pipe. There are two ways of doing this, as shown in the drawings. Protect your eyes by wearing safety glasses or goggles. If the pipes are joined by flexible connectors or unions, simply unscrew them (A). If there are no connectors, you must cut through the pipe with a hacksaw (B).

4 Remove the power line by one of the two methods shown here. If the water heater is supplied by gas, check to make sure that the gas is turned off and then disconnect the supply pipe with a wrench (A). Remove the draft diverter from the flue at the top of the heater. Check the flue for proper draft with a match. Sometimes flues are filled with dead birds, cement, or other debris and need to be cleaned before the new draft diverter is connected. If you have an electric-powered heater, first check to make sure that the circuit is shut down and that there is no incoming power. Then disconnect the wires at the incoming electrical cable. This is usually located at the top of the tank (B).

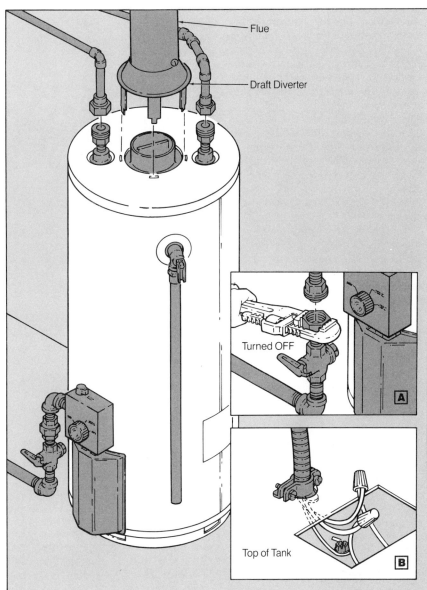

Flue

Draft Diverter

Turned OFF

Top of Tank

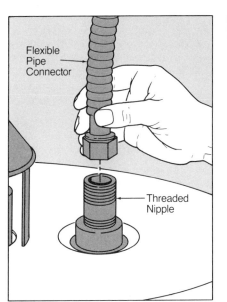

Flexible Pipe Connector

Threaded Nipple

7 Hook up the plumbing to the new heater. If the old piping doesn't quite meet the new fittings, use flexible pipe connectors or unions to connect the water and gas lines. These lengths of flexible tubing are simply screwed onto the pipe and bent for the hookup. If your pipes aren't already threaded, measure and cut for threaded nipples. Install the nipples and then attach the flexible pipe connectors with a wrench.

Activate a new electric heater by first hooking up the electrical wiring (the circuit should still be shut down). With all plumbing connections made, open the cold water inlet valve. Once the tank is filled with water, open up the hot water faucets to 'bleed' the pipes. Squeeze the lever of the temperature and pressure-relief valve to test it. Restore power by either opening the gas inlet valve, inserting the correct fuse, or switching the circuit breaker on.

Gas heaters need to have the pilot relit; follow instructions on the control panel for this step. Set the water temperature as desired.

Check all gas connections by applying soapy water with a brush at the connection; wait awhile to see if bubbles appear. The appearance of bubbles indicates a leak; continue to tighten the connection and test it until the problem is corrected.

WARNING

If you need to run new gas piping for a water heater installation, you should call a professional; this task is too dangerous for the basic handyman. Also, when installing an electric heater, you should use extra caution when making the electrical hookup and know exactly what you are doing.

Water Heater Safety Measures. There are two parts of a water heater that are crucial for your own safety as well as a long life for the fixture. The most important of these is the *temperature and pressure-relief valve.* This valve relieves pressure and prevents a steam explosion if the thermostat malfunctions and the pressure or temperature exceeds abnormal limits. Most water heaters come equipped with such valves and if they do not, they will have a fitting at the top or side of the tank for you to install one. Don't ever try to salvage an old temperature and pressure-relief valve, and when purchasing a new one, make sure that it is compatible with your water heater.

The other safety component that is considered essential is the overflow pipe that connects to the temperature and pressure-relief valve. This long, vertical pipe protects you from being scalded when you check the valve—which you should do on a regular basis.

Maintenance of Your Water Heater. You should check your temperature and pressure-relief valve every three months. This is a very easy process; all you need is a few minutes and a bucket. Put the bucket under the overflow pipe and open the valve. If hot water does not empty into the bucket, you should replace the valve.

Another routine maintenance involves the emptying of the tank of sediments. This can be very important depending on what kind of minerals are in your local water supply and also what the tank is composed of; metal-lined tanks will rust faster than glass-lined tanks. When sediment builds up, the heating efficiency of the tank is reduced. So clearing the tank every once in a while will give you the most for your money spent on energy.

Also, if your tank has been making noises, it's liable to be caked with sediment. What you're hearing in this case is actually a series of little steam explosions. Water, trapped between layers of sediment, is seeking an escape and is signalling that your tank is now in need of a thorough flushing. In some cases, you can use a cleaning compound to get rid of the accumulation, but be sure to follow the manufacturer's instructions precisely.

Help for Impure Water— A Replaceable Filter

If your water has an undesirable taste or odor you might want to have it tested to find out what chemical or mineral is causing the problem (page 5). Health

INSTALLING AN OVERFLOW PIPE

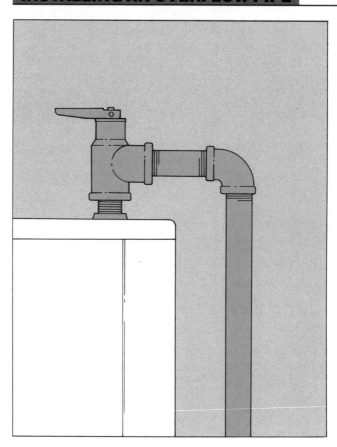

Purchase two pieces of pre-threaded steel pipe the same diameter as the outlet of the temperature and pressure-relief valve. One piece should be long enough to extend just beyond the side of the water heater. The other, longer piece should be of a length to extend from that point to within 6 inches of the floor. With two wrenches, attach the short pipe to the outlet. Connect an elbow to it and then thread the longer pipe to the bottom end of the elbow.

problems resulting from impure water are becoming increasingly common, so it's to your advantage to stay posted about the condition of your local water. Other annoyances include soap scum around the rim of the bathtub and deposits in pipes, water heaters, and clothes washers. Depending on the severity of the problem, the installation of a water softener or chemical injector might be warranted.

An inexpensive but fairly effective alternative that can be installed by the do-it-yourselfer is a replaceable water filter. These small devices have replaceable charcoal or carbon cores in their 'tanks' which trap impurities as water flows through them. They can be installed in the main line, past the water meter, to filter all of the water coming into the home, or they can be installed on separate cold-water lines, such as at a kitchen sink. In the latter case, only cooking and drinking water would be filtered.

Often, replaceable water filters come in kits but even if you need to assemble parts yourself, the installation will be easy. The most important factor is the position of the filter; it must be upright to operate correctly. Flowing

water must be halted so shutoff valves are placed on either side of the filter. Installing a filter in a horizontal line requires correct measuring and cutting of pipe and the use of compression fittings. Putting a filter in a vertical supply line is a bit more complex, requiring that you first reroute the pipe in a loop.

Regular replacements of the core should be made every six months to a year, depending on consumption and the water's condition. This procedure is very easy; simply turn off the valves and with a bucket underneath, unscrew the filter body from the cap. Remove the old filter, insert the new one, and re-screw the filter body back into place.

Getting Rid of Water with Sump Pumps

Depending on where you live, your home might have a basement that stays wet all the time from groundwater seepage. Another common scenario is to see tiny streams forming after three or four days of rain. If either is the case, you probably have experienced mustiness and mildew in this area of your home. A remedy for this problem is to install an automatic sump pump.

Sump pumps evacuate water from a pit or 'sump' which is actually just a hole in your basement floor. When the water level in one of these devices reaches a preset level, a switch turns on the pump. A special valve, called a *check valve*, additionally prevents water from draining back into the sump when the pump stops. Sump pumps are available in two basic types: submersible and the pedestal-type.

The pedestal-type sump pump consists basically of an electric motor on top of a pedestal. There must always be weter in the pit as a pump can be damaged if it becomes dry. The submersible type is more expensive but is often preferred for its features. In requires less maintenance and cannot be damaged as easily as the pedestal-type.

If seepage or rainwater is causing the wetness, the sump pit should be located at the lowest point in the basement. If wetness is caused by a draining sink or washer, the pump can be installed anywhere that it's convenient to run a drainpipe. Installation basically involves preparing the pit which will have a concrete bottom and sides made of concrete or concrete and terra cotta. Once the concrete is set you will hook the pipe up to your sewer line or seepage pit; unions and check valves are required by most codes. Follow the manufacturer's instructions completely, especially when making the electrical hookup. If you need to make your own cover for the pump, make sure that it cannot be removed by children.

A Toilet in the Basement

Though homes aren't often designed with toilets in their basements, the addition of one can be a real convenience, especially if the area is used often for work or play. In certain regions of the Northeast United States, this fixture can be installed singly, without the aid of a pump to force the waste upward. In these situations, which should be checked thoroughly by the local code, the water pressure must be high—approximately 40 psi or more. Such *up-flushing toilets* are constructed no differently than regular toilets and require the same basic steps for installation. In some cases, also, the level of your street sewer will be lower than your basement level so that your basement toilet will not be up-flushing.

Usually, however, local codes will prohibit the up-flushing toilet. Instead, they will recommend an alternative that consists of a standard toilet supplemented with a sewage ejector to get rid of the waste. *Sewage ejectors* are merely sealed tanks with pumps inside of them. Expensive installations, these require excavating below your basement floor for both the tank and the 4-inch drainpipe from the toilet to the tank. The materials and the labor take such a project out of the range of many homeowners, but check your area for prices.

CAUTION

Before installing a water filter, turn off the water at the nearest shutoff valve or the main shutoff valve (page 9). Then drain the pipe.

Outdoor Plumbing

Once you've made plumbing repairs and improvements within the confines of your home, it's time to open the doors and take a critical look at the rest of your property. There are many time- and energy-saving improvements to be made outdoors also.

You can keep your lawn greener by installing an outdoor spigot. Such an installation saves you the expense and the bother of dragging a long hose halfway across the yard. This project is detailed plus information is given on freeze-proof faucets—a great device for people living in colder climates.

If you have a garden or you're a stickler for watering the lawn, you can go all out and install an underground sprinkler system. This is truly not within the realm of the home do-it-yourselfer but some tips are given to help you deal with companies that install sprinklers. If you live in a warm climate or have a beach house, you might want an outdoor shower.

For people who live in rural areas, the subject of septic tanks is discussed and, related to that, drainage fields. Outdoor plumbing requires that you check local codes first and pay special attention to safety rules.

General Rules for Outdoor Pipe Runs

Your locality is the first factor that you should consider when running pipe outside of your home. It will affect both how the pipes are run and what kind of fixtures you should install. Other important factors are the house structure and the present landscaping. In areas where winters are cold, any outdoor system, such as for lawn sprinkling, must be designed to permit complete drainage of all portions above the frost line.

If the house has a basement, this can usually be acomplished by pitching the main pipe downward toward its point of entry through the basement wall. It is then possible to drain the system through a valve inside the house (see page 117 for installation instruc-

tions). If the house has no basement or if the lawn slopes away from the house, the pipe will instead be pitched downward away from the house to a small pit about a foot or two deeper than where the pipe ends. With an ordinary sillcock or an automatic drain valve on the pit end of the pipe, it can be drained into the pit. The pit need not be much more than a shovel-width square, as pipe of the size commonly used (about ¾ inch) doesn't hold a large amount of water in normal-length runs.

As far as materials, the most convenient type of piping is polyethylene (PE) but, as always, be sure to check your local code before assuming that this is acceptable. Some codes have restrictions about tying PE pipe into the indoor service line. This type of plastic pipe is

especially useful because of its flexibility; it can be bent into gentle curves, a real benefit when you meet with obstructions as sometimes happens outdoors. It's also easy to use because it can be joined with simple clamps.

Fittings tend to reduce water pressure slightly so it is preferable not to use many of them. If you must do so, such as when you put in a sprinkler system, then you should use PVC (polyvinyl-chloride) pipe. It's not as flexible as polyethylene but the strong cemented fittings will assure you less chance of leaks.

The size of the outdoor pipe depends on the pipe size at the source. This is almost always ¾ inch but if it is smaller or larger, match your new run to the existing pipe.

For Dry Lawns or Dirty Cars—Add a New Faucet Outside

Most homes are built with at least one faucet outdoors but it isn't always where you need it. You might have to buy extra-long pieces of hose to get water to your flower beds—and then you have the aggravation of wrapping and unwrapping the hose everytime you need to use it. A new faucet, or *hose bibb*, outside your home can save you all this bother when doing your outdoor chores. Installation is fairly straightforward as long as your plumbing skills are down pat (pages 19-33).

First, you'll need to do some planning. The procedure shown here is with the ideal situation; the faucet's position above the foundation will be a good height for proper use. If your home is built differently and your foundation is quite low, however, the height of your new faucet might not be suitable. Perhaps you'll be able to fit a hose onto the faucet by stooping over, but you won't be able to fit a watering can underneath it.

An option in this case that will provide you with a higher faucet, is to locate the faucet on the outside wall of an attached garage. If the pipe is run from here to an adjacent house wall, it can then be run to the basement for tying into the supply line. Other options might be available; you'll need to study your system to locate the perfect route.

The fixture itself, the hose bibb, comes in several varieties. Most types have a threaded base and a threaded spout for attaching a garden hose. One popular type, called a *sillcock*, comes with a flange or escutcheon. This feature makes for an easy and secure attachment to the exterior wall of your house; it's held in place by screws.

The *freezeproof sillcock* is recommended if you live in an area where winter temperatures take more than an occasional plunge below the freezing point. These elongated fixtures actually stop the water flow before it reaches outside and they also drain themselves if installed correctly—on a downward slope to the ground outside. If you do not use a freezeproof sillcock in a cold climate, you should install a valve in the basement for draining the faucet during the winter.

Once you have installed your new sillcock or faucet, you might want to equip it with a preventive accessory—an *anti-siphon device*. Also called a *vacuum breaker*, this small device prevents the backflow of used or otherwise pol-

INSTALLING AN OUTDOOR FAUCET

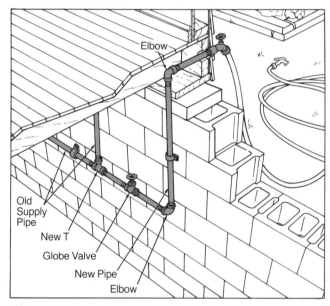

1 The ideal location for the outdoor faucet, or sillcock, is on a wood wall of the house, just above the foundation. First, inspect the proposed area from the basement to make sure that there are no obstructions such as studs. Also, note the location of the nearest cold-water pipe from where you will run your supply pipe. Then, with a spade bit, drill a hole just large enough to accept the pipe. If you need to make the hole in masonry, use a masonry bit and work at a horizontal strip of mortar or at the hollow part of a concrete block. (Older three-core blocks are hollow in the center; newer two-core blocks are reinforced in the center.) Determine what kind of faucet to install. If you live in a climate with cold winters, consider a freeze-proof faucet (right); they come preassembled with long faucet bodies and they should be installed at a slight downward slope from inside to outdoors. If you don't use the freeze-proof type, then you must run piping for the faucet through the wall.

2 First, turn off the water at the main shutoff valve (page 9) and empty all pipes to prepare for tapping into the cold-water pipe. In this illustration the supply pipe formerly made a right-angle turn upstairs. In place of this elbow, insert a T fitting. You can use a saddle tee or a soldered fitting. From it run a short length of piping, a globe valve, and another short length of piping. Using two elbows, run pipe up to the entrance for the faucet. If necessary, use an adapter when connecting the stem of a freezeproof faucet to the elbow. Run a length of pipe through the wall and connect the new faucet. Anchor the pipes to the studs and joists near the wall and along the run. Fill gaps around the pipe with waterproof silicone caulking. For a flanged faucet, apply caulking around the pipe before screwing it down; when the flange is tightened, a seal will be formed.

luted water. Often used indoors also, these help to keep your house supply clean. Simply thread them into the faucet spout.

Longer Outdoor Runs

As mentioned earlier, you can extend pipe away from your house as long as a draining slope is provided. There are many options for the end of the run; you can place freestanding faucets, showers or other fixtures here. As for the beginning of the run, you can tap directly into the house supply line or you may begin at an existing outdoor faucet.

When the line needs draining in the winter, this can be done outdoors at a drainage pit or indoors at a newly installed *stop-and-waste* valve (below). This valve, somewhat like a globe valve (page 60), is constructed so that you can stop the water flow from the supply line while emptying the pipe further up the line (in this case, the outdoor pipe).

If you choose to add a drainage pit, install an *automatic drain valve* at the end of the run. The automatic drain valve (page 119) is constructed with a loose ball in its central chamber. When water is turned on, the pressure causes the ball to close up the drain hole and the water goes out the nearby outlet. When water is turned off the ball rolls back into its chamber allowing the remaining water in the pipe to be drained into the drainage pit.

Instructions are given for an outdoor run from a faucet with the choice of adding a drainage pit or a stop-and-waste valve in the basement. Study them very carefully before making your plan.

Erecting a Lawn Hydrant. Especially if you have a permanent garden space, you might want to install a short free-standing faucet, or hydrant, in your backyard. This is a relatively easy installation, requiring only a few special materials and little labor. Note, however, that though the hydrant on page 120 is at the end of a run from a basement where a valve is placed (with no outdoor drainage necessary), a hydrant can also be installed with a drainage pit at the end of a run. As an option, too, it can be installed in the middle of a run.

Watering Without the Hose—Sprinkler Systems

Underground sprinkler systems can be a worthy investment for the homeowner, especially if the lawn is large, the climate is moderate, and much time is spent watering. Often homeowners choose to have such systems installed

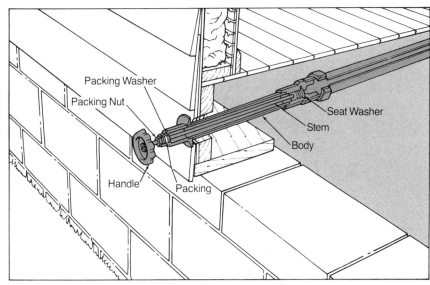

Anatomy of a Freezeproof Faucet. Highly recommended in geographical areas where the temperature often dips below freezing, this faucet is designed to prevent freezing pipes. Basically, it is like a compression faucet but it has an extra-long body and the valve seat is located far back from the handle. Installed, the valve seat is within the confines of the home, safe from freezing temperatures. Such a faucet must be installed at a slight downward slope so that the body is emptied of all water when it is shut off. This type of faucet may not be tapped for an outdoor extension.

A common error in using this faucet is to turn the handle too tightly. When the water is turned off, it naturally takes a longer period of time for the long body to drain than it does when a regular faucet is used. Unknowingly, many people turn the handle much too tightly and thus quickly wear out the seat washer.

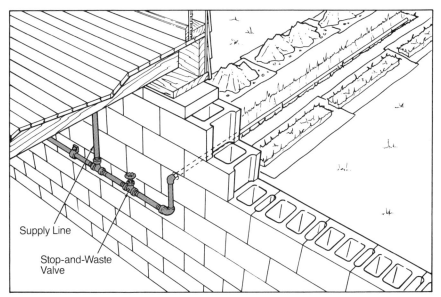

An Option for Draining Outdoor Pipes. This installation is relatively easy because it saves you digging a drainage pit. In order to drain the outdoor piping for the winter, you will simply open up the stop-and-waste valve and drain the line. The installation is only feasible, however, if the ground outside your house is level or slopes downward toward the house. Correspondingly, pipe will be run at a downward slope toward the house and through the wall. The type pipe you use will be determined by local code and what fixture is being installed outdoors. Common choices are PE or PVC, both being easy to work with. Begin by digging the trench for the outdoor pipe (page 119) and boring a hole through the house wall at the end of the trench (page 116) using a slight downward pitch for both steps.

In the basement depicted here, there was formerly an elbow in the supply line. At that point, install a T and then a short length of pipe. To it, attach a *stop-and-waste* valve, similar to a globe valve (page 60), but containing a tap for draining water from the run ahead of it. If necessary, use adapters to connect different kinds of pipe or fittings. The valve has an arrow on it that must be pointed in the direction of the water flow. From the valve, run piping through the wall, outdoors and to the end of the trench, again with a downward pitch toward the house. Install an elbow at the end of the run for the fixture connection.

RUNNING PIPE FROM AN OUTDOOR FAUCET

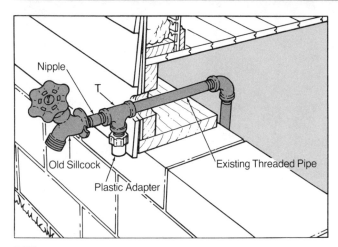

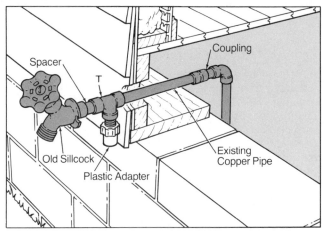

1 First, install a T fitting at the pipe for the existing sillcock. This is done differently depending on whether the piping is threaded or copper. If the pipe is threaded, remove the screw that holds the sillcock to the wall. Then, using a 10-inch pipe wrench, unscrew the sillcock. Screw a T fitting that includes a tapped inlet onto the pipe; tighten it until the inlet faces straight downward. Extend the T horizontally by screwing in a nipple and then adding the old sillcock. Into the T inlet, screw a plastic adapter for the new run.

If the pipe is copper, cut it in the basement, at least 2 inches from the inside wall. Detach the sillcock by removing any screws holding it in place; then grip it from the outside and pull it and its attached pipe out of the wall. Wearing eye and hand protection and taking fire precautions, separate the sillcock from the pipe with a propane torch (page 24). Then sweat a T with a tapped inlet onto the salvaged section of pipe; follow it with a short spacer and the old sillcock. Position the assembly so that the inlet will point downward while the faucet handle faces upward. Push this assembly back into the hole and join it to the cut pipe with a coupling. Into the T inlet, screw a plastic adapter for the new run.

4 Lay all of the pipe in the ditch along the run. Starting at the house, prop up the pipe with bricks or stones at 6-foot intervals to create a slight downward pitch toward the end of the run. Periodically, check the pitch with a carpenter's level. Place the level on top of the pipe at intervals and rearrange pipe until the bubble in the level is always slightly to the house side of center. Once the pipe is correctly positioned, use loose soil to shore it up between stones.

CAUTION

Before tapping the water supply in the basement, turn off the water at the main shutoff valve (page 9). Then open nearby faucets to drain the line as much as possible.

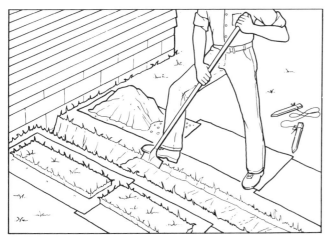

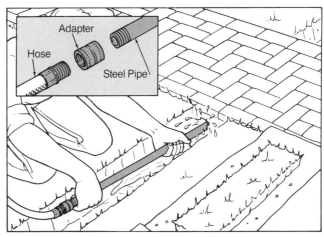

2 At the point where you wish to install your new outdoor fixture, drive a stake. Drive another stake at the house piping exit. Extend a string from stake to stake. Mark for a 6-inch wide trench by inserting a flat shovel at 3 inches on both sides of the string. Make connecting grooves, about 2 inches deep, along the length of the trench. Remove the stakes and string and make grooves at the ends to form a long rectangle. Spread large plastic sheets or tarpaulins on the ground, some for loose soil and others for sod. To make sod, use the shovel to divide the long rectangle into smaller segments, approximately 5 feet long. Then push the shovel underneath and work it up and down to free the grass roots. Continue until an entire segment is loose, pick up the sod carpet and lay it on the plastic sheet. Once all the sod is removed, continue to dig from the sides to create a V-shaped trench with the low point measuring about 10 inches deep. Put the loose soil on separate plastic sheeting.

3 If your run must go under a masonry sidewalk you can tunnel under it with a few special accessories. Purchase an adapter for connecting a regular garden hose with a ¾-inch threaded steel pipe. Also order the 3-foot-long pipe, threaded at one end for the connection. Screw it all together tightly and then turn on the water full force. Quickly insert the 'water pick' into the ground and push the pipe forward gradually until it reaches about halfway under the sidewalk. Repeat the process working from the other side.

5 At the end of the run, dig a pit 12 inches square and about 20 inches deep. Fill it with coarse gravel level to the bottom of the adjacent trench. At the pipe end, install a T with the inlet pointing up. To the T's open arm add a pipe about 6 inches long—run it to the center of the pit.

6 To the end of the pipe over the pit, install a threaded adapter. To it, attach an automatic drain valve. Cut a piece of galvanized pipe for it—slightly larger in diameter and twice as long as the valve. Add 3 more inches of gravel to the pit. (See page 117 for explanation of the drain valve.)

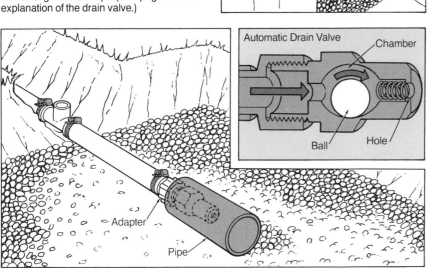

7 Last, connect the pipe to the sillcock. In the trench, attach an elbow onto the pipe, facing upward. Measure the distance from the plastic adapter at the sillcock to the elbow. Install a length of pipe between them. To protect pipes, apply two coats of latex paint to the above-ground plastic piping. Depending on what you are installing at the new outlet, install risers in the T at the end of the run. Make the outlet connections as needed. Turn the water on and test the line for leaks. Shovel dirt into the trench and the pit, about 2 inches higher than the pipe. Hose down the dirt and wait one day. Tamp it with a 2 x 4, fill in the trench and pit and replace sod.

by a lawn-watering service. If you are considering adopting such a service, you should definitely read the fine print to understand what you're getting. As a worse-case scenario, a company will tap into a water line independent of your supply system so that your fee will not be part of your regular utility bill. You might have to pay a price much higher than what you pay for your house water and the charge could also be made in months when you aren't even using the service!

If you connect your system to your regular water supply, then you will pay no more for it than you do for the house water. An important factor in this arrangement, however, is the water pressure. Since your system is liable to have several outlets run in long lengths, your house pressure will be significantly reduced. In fact, your current water pressure will determine how large your sprinkler system can be.

Lawn sprinkler systems come in many forms. The simplest kind will have a manual control with each section of the lawn on a different line and controlled by a globe or gate valve. Automatically controlled systems are more expensive but they perform the starting and stopping for you. The advantage here is that the system can operate at times when you aren't at home or at hours when other home fixtures aren't being used. There are several types available. One has buried boxes that contain the electrical wiring and wiring components plus a shutoff valve. Another type has a control valve above the ground, usually close to the house, to which an electric timer is attached. Adjacent to the control valve will be an antisiphon valve. The function of this fixture is to prevent the backflow of water to the house supply. Antisiphon valves are necessary, even critical, for systems that will be used for spreading lawn chemicals. The backflow of certain chemicals in supply systems can be fatal.

Materials vary but plastic is universally accepted as the best kind of piping to use for the underground runs; it will expand rather than crack in low temperatures. Rigid PVC or PE may be chosen depending on your code recommendations. Risers, the vertical pipes on which the sprinkler heads are placed, may be threaded metal or plastic. Heads can be composed of brass, a metal highly resistant to corrosion, or plastic. When different materials are used, easy connections are made with adapters.

Pop-up type sprinkler heads are the standard. These rest close to the ground when not in use so that you can mow your lawn without fear of striking a fixture. When the system is turned on they rise to the necessary height to spray water over a given area. They also may be raised up by hand for regular cleanings.

Planning for a System. The use of plastic pipe has lowered the cost of this project; still it is a project not recommended for the beginning home plumber. Instead, you should seek the help of experts. To begin the planning process, make a rough sketch of your lawn indicating which areas you want watered. Take this to a plumbing supply store or, preferably, a company that specializes in sprinkling systems. In the drawing, also include the location of the house and any walkways, fences, trees, and permanent beds or gardens. It's also helpful to bring a sketch of the house since the pipe might be run from inside. Other specific details that your supplier might need include: the condition of the soil, the type and diameter of the supply pipe that you will tap into, and the inside diameter of your water meter. This latter information might be printed on the meter housing; if it isn't, find out by calling your water company.

As mentioned before, the water pressure is critical. Your supplier will measure your pressure with a special gauge made for this purpose. Once you know your pressure, a designer will make a detailed plan for you indicating

INSTALLING A LAWN HYDRANT

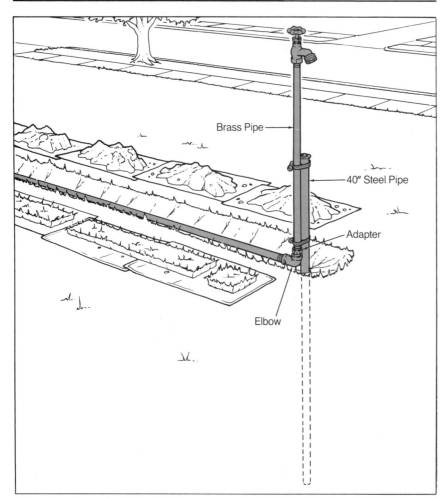

Brass Pipe

40" Steel Pipe

Adapter

Elbow

Note: Though this illustration shows an installation at the end of a run from a basement (with no outdoor drainage necessary), the hydrant can also be installed next to a drainage pit at the end of a run. Also, it can be installed in the middle of a run. To begin, cut a support pipe for the hydrant. It should be ¾-inch steel and measure 40 inches long. Next to the elbow or T in the trench, set this steel pipe and then drive it in with a sledgehammer to a depth of about 24 inches. Purchase a brass pipe as the riser for the hydrant. It should be threaded at both ends and should measure about 8 inches longer than the preferred height of your new hydrant. If necessary, attach an adapter to the elbow or T in the trench and then install the brass pipe on it. Secure the brass and steel pipes together using stainless steel pipe clamps. Turn the water on briefly to flush out dirty water. Then, screw the hydrant onto the brass pipe, turn the water on again and check for leaks.

the locations of all sprinklers. The design service is usually provided free of charge, as long as you buy the dealer's fixtures and services. Different spray patterns will be created depending on how they're placed. Sprinklers are usually positioned in groups with each group having its own control valve.

Soil and Sand Restraints— Outdoor Showers

Outdoor showers are not much different than other types of outdoor plumbing projects. You must, however, be sure to provide for adquate draining or else you will have trouble with wetness close to your house or wherever you install the shower. This involves digging a trench and running drainpipe at a slight downward pitch to a spot away from the shower drain.

The supply pipe will be run from inside your house and up an outside wall; it is supported on the wall by clamps and furring strips. The shower head is installed at the top and a valve is installed at a height convenient for whomever will use the fixture. Drainpipe goes to a *dry well*, a deep gravel-filled pit a good distance away from the house and other drainage fields. When you put in the base, you should follow the manufacturer's instructions.

Instead of installing the shower against the side of the house, you can use the same basic methods to install a freestanding shower. The only difference is that the vertical supply pipe will have to be strengthened. The usual method is to use 2-inch steel pipe which is driven about a foot into the ground and rises to the height of the shutoff valve. The water supply pipe is simply inserted inside of this pipe—so its function is strictly that of a support piece.

Private Systems—Septic Tanks and Drainage Fields

If you live in a rural area, your home plumbing might be hooked up to a septic tank, an amazingly simple invention for disposing of home sewage. Though installation of septic tanks, a job way beyond the realm of the amateur, will not be covered here, some basic information is presented to assist you when purchasing such a system or maintaining one.

How Septic Tanks Work. In operation, a sealed-joint pipe carries all sewage and other plumbing drainage from your house to the septic tank. The inlet is slightly higher than the outlet, so

for each amount of fluid that enters, an equal amount flows out into the absorption field. To prevent the incoming sewage from flowing directly out the overflow, both the inlet and outlet are fitted with down-pointing sanitary Ts or shielded by baffle plates. This way, the incoming flow is directed downward to the bottom of the tank, while the outgoing flow is drained off from the top of the tank.

Bacterial action takes place at all levels in the tank, breaking down the solids to liquid, gas, and mineral sludge. The liquid is what flows into the drainage field. The gas drifts back up the sealed-joint pipe to the house and then escapes up the vent stack and into the open air above the house. The sludge settles to the bottom of the tank—and this is what occasionally must be pumped out by firms operating trucks designed for the work. The sludge accumulates very slowly compared to the amount of solid matter and liquid entering the tank.

Properly maintained, a septic tank should be trouble-free for a long time. Many have required no attention for ten years or more, and then only a professional pumping out. Even so, tanks should be checked at least once a year by a qualified inspector. The best time to have a tank pumped is in the spring. In cold winters, solid waste cannot be broken down easily, thus slowing down the bacterial action; therefore pumping in the autumn is not as beneficial.

Problems Begin When… Many people don't understand the importance that bacteria plays in this situation so they unwittingly treat their fixtures and pipes just like those in a public waste-disposal system. Household chemicals retard the necessary bacterial action by killing considerable amounts of bacteria. If the chemical inflow is not excessive, however, the tank continues to perform effectively.

Large-capacity automatic washing machines and similar tank-overloading appliances can further complicate the problem. By sending large volumes of water into the tank each time they empty, they churn up the solids that are still being broken down by bacterial action—and they also churn up the sludge. When these products flow into the drainage field, they clog the pores of the soil and reduce its rate of absorption. In severe cases, the liquid effluent from the tank, not being absorbed by the earth, finds its way to the surface and overflows onto the lawn. As this can easily be prevented in most cases, the homeowner, not the tank, is to blame.

Precautions for Septic Tank Use. Here is a checklist of important warnings about septic tanks:

■ Don't ever dispose of chemicals through this system. They often retard the bacteria that is necessary for attacking and disintegrating solid wastes in the tank.

■ Never use chemical cleaners, such as those for unclogging drains, for the same aforementioned reason.

■ Don't ever dispose of thick paper products even if the advertiser claims that you can. These clog the main drain to the tank and also the smaller pipes to the dispersal field, shutting down the entire system.

Solving Drainage Problems. There are some steps that you can take to solve or avert drainage problems. You can switch to appliances that are designed to economize on water.

If you are already using water-saving appliances but still have effluent overflow, you may have enough property to separate your sewage (toilet) system from your other drainage. This is done by diverting the drainage from all fixtures and appliances *except toilets* to a separate main drain. The drain may be led to dry wells or *drainage pits*. These are pits walled with uncemented masonry blocks to prevent the sides from caving in. As only a slight amount of solid matter is carried by the *gray water* from sinks, wash basins, tubs, and washing machines, the fluid in the pit has little tendency to clog the spaces between the masonry blocks.

If there is an accessible place in the house for a 'grease trap', it may be worthwhile to install one. Most of the grease in the drainage comes from the kitchen sink. Over a period of time it coats the walls of the dry well, clogs the soil, and slows absorption. To be useful, the grease trap must be where it can be cleaned easily.

A giant-sized septic tank is another measure often used to minimize field-clogging problems. Yet another 'last resort' is to take your wash to a laundry. The corrective or preventive measures you use may depend on your local code. Some codes require that all drainage from all fixtures be piped to the septic tank. Others recommend that sewage from toilets and drainage from other fixtures be handled separately. So check your local code before you plan a new system or modify an existing one.

APPENDIX 1: READING YOUR WATER METER

There are three common types of water meters—none of which is difficult to read. Taking a periodic check of your water usage will help you to resolve any billing disputes with your water company plus it will aid you in detecting leaks in your system.

Six-Dial Meter. This is the most popular device for residential use. Five of its six dials measure the number of cubic feet of water used for one revolution. Each of these dials (labeled 10, 100, 1,000, 10,000, and 100,000) are divided into tenths. The remaining dial, not divided, measures a single cubic foot per revolution.

The only tricky part is the way in which the dials are numbered and the direction in which the needle moves. The needles of the 100 and 10,000 meters move clockwise and the needles of the other three dials move counterclockwise. When you read a six-dial meter, you should begin with the dial labeled 100,000, and note the smaller of the two numbers nearest the needle. Next, read the dial labeled 10,000 and then continue on reading the 'smaller' dials.

Five-Dial Meter. A five-dial meter is read in exactly the same way as a six-dial meter except that there is no single cubic foot 'clock'. This measurement is taken instead by reading the large sweeping needle that moves across the face of the entire meter.

Digital Meter. Obviously the easiest to read, this type is like an odometer. At a glance, it gives you the total number of cubic feet of water consumed. Some models also have a small dial that measures a single cubic foot of water per revolution.

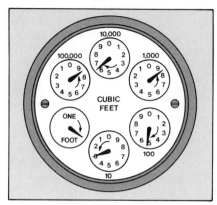

This six-dial meter reads 868,530.

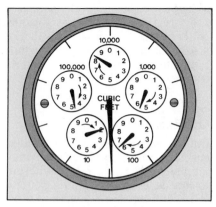

This five-dial meter reads 485,620.

To detect a leak, turn off all water-using appliances and note the single cubic foot dial. If it moves at all, even slowly, you probably have a leak in your system. To determine how much water it takes for a certain task such as filling a swimming pool, a waterbed or watering your lawn, simply take a reading beforehand; take one after, and then subtract the difference. To convert cubic feet to gallons, multiply the figure by 7.5, the number of gallons in a cubic foot.

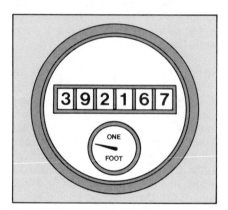

This digital meter reads 392,167.

APPENDIX 2: TROUBLESHOOTING HOME PLUMBING

PROBLEM	POSSIBLE CAUSE	SOLUTION
Not enough water.	Low water pressure.	Notify your water company. If your service checks out as adequate, your pipes might contain lime or sediment. Consider replacing supply pipes.
		Install a pump.
Too much water.	High water pressure.	Notify your water company and/or local government officials.
		Install a pressure-reducing valve.
Water tastes or looks bad.	Impure water.	See the guide on page 4 to diagnose and solve the problem.
Noise in the system.	Water hammer caused by a valve being shut off at an appliance, such as a clothes washer.	Install air chambers at the fixture.
	Pipes improperly fastened to house framing.	Refasten pipes or add wooden wedges to secure pipe.
	Toilet needs repair.	See the guide on page 52 to diagnose and solve the problem.
	Partly opened valve.	Shut valve completely off.
	Bad washer in a stem faucet.	Replace washer.
	Gurgling sound from faulty or plugged venting.	Clean vent.
		Install an anti-siphon trap under fixture.

APPENDIX 3:   HOW TO REDUCE YOUR WATER BILLS

Reducing your water bill is a matter of making repairs when they're needed, changing your water-use habits, and installing water-conserving fixtures. Use the guides throughout this book to detect and repair all leaks, both in the lines and at fixtures. Make repairs as soon as possible.

Change your habits by turning off all faucets completely and educating all family members to do the same. Other measures include rinsing off dishes and other objects in basins instead of with a running faucet and also running the dishwasher only when it is full. Be sure also to use appropriate settings on your clothes washer for the size load of laundry that you're washing; this appliance is a particularly high-volume water-guzzler. Take care to use less water in the bathroom where a typical family's usage measures ¾ of its total. Take shallower baths and shorter showers and use a public facility's, such as a school's or a health spa's shower whenever possible.

Water-Conserving Devices.   Aerators at faucets deliver splash-free streams and they're inexpensive and easy to install. Similarly, a *faucet flow controller* will reduce the water flow from approximately 5 to 8 gallons

to 4 gallons per minute but the spray force will still be maintained. Some new faucets come already equipped with such devices.

For your shower, similar devices are available. These are usually inserted between the supply pipe and the shower head. Again, many new shower heads come equipped with this feature. The flow of water can be reduced from about 5 to 8 gallons per minute to as few as 3 gallons per minute. Water will be conserved and since it is primarily hot water, other forms of home energy will be likewise saved. Another conservation fixture, as well as a time-saver, is a *mixing valve*. These thermostatic devices control and keep your preferred water temperature ready for you, saving you the chore and the water spent every time you mix it for bathing.

Toilets consume a lot of water and if you're purchasing a new one you should consider a water-conserving model. The ultimate water-saver is a pressure-valve type toilet, often seen in commercial buildings. Unfortunately for the homeowner, these require a 1-inch supply pipe instead of the usual ½-inch pipe, so a replacement is usually not feasible. However, if you're building a new home, you should consider plumbing for such a fixture.

Other options are to install dams within your toilet's tank or to add displacement fixtures, such as plastic bottles filled with water. Dams are simply rectangular pieces of plastic or rubber-coated metal. The average toilet uses 5 to 7 gallons per flush and both of these measures can save anywhere from ½ to 2 gallons of water every time it is flushed.

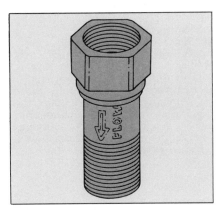

Faucet flow controller.

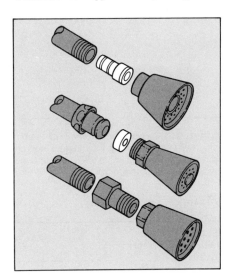

Inserts for shower flow control.

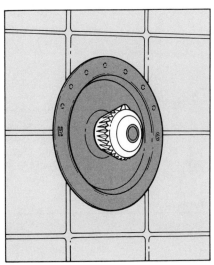

Mixing valve.

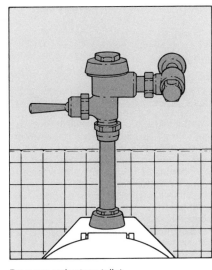

Pressure-valve type toilet.

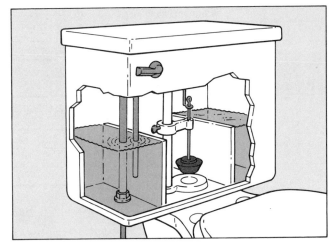

Toilet dams.

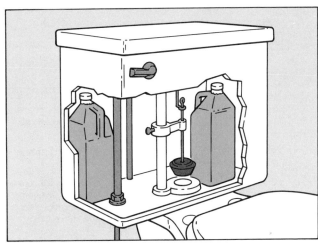

Plastic bottles.

APPENDIX 4:   SHUTTING DOWN A PLUMBING SYSTEM

Use these steps to completely shut down a vacated home for winter months.

■ First, turn off the main shutoff valve or have the entire service stopped for the house by calling the water company.

■ Then, starting at the home's top floor, open all faucets and valves completely. Be sure to open up all faucets outside also.

■ When the pipes are completely cleared, open the plug at the main shutoff valve and let it drain. (This step might require a call to your water company.)

■ Shut down the power to your water heater (page 110) and let it drain completely.

■ Flush toilets and pour a gallon of automotive antifreeze (mixed with water) into each toilet bowl, the tanks, and all sink traps. If your house has a main house trap or floor drain, fill it with full-strength automotive antifreeze. In short, antifreeze or an antifreeze mixture should be added anywhere water is not potable.

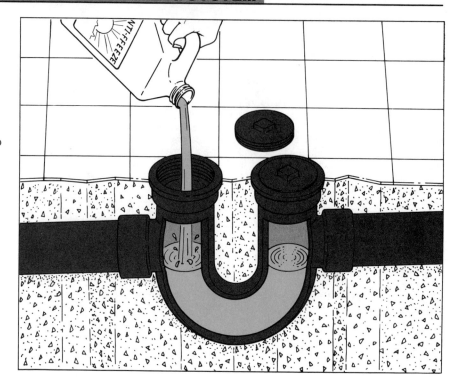

Pour full-strength antifreeze in the house trap or basement floor drain.

APPENDIX 5:   BUDGETING PLUMBING RUNS

Contractors or plumbers would draw a schematic line drawing for a small run with only one or two turns in it. For larger jobs with many branches, they would make an isometric drawing of the system they wanted to price. Isometrics are three-dimensional line drawings showing all turns, branches, and fittings on the system. Another way to count the fittings on a pipe run, and an easier method for the beginner, is to draw two schematics—one looking down from above and one looking from the side. Count all the fittings on one drawing and then check the other drawing to see if there are any hidden fittings that you didn't see in the first drawing.

Here are the materials and pricing methods for the project on page 240, installing an outdoor faucet:

2, ½-inch 90° elbows X $.50 each = $1.00
30', ½-inch M copper tubing X $.48 per foot = $14.40
1, ½-inch stop and waste valve = $4.80
1, ½-inch X 12″ freezeproof sillcock = $14.20

These are estimates only and will vary according to locale; they are presented only to illustrate the method used. When budgeting for your plumbing run, be sure to fill out a budgeting sheet, such as the one shown here, and try to incorporate all materials, including needed tools. To price materials after the list is made, you can call a plumbing supply house, tell them what you need and write in their price. You might want to compare prices if you're making long runs or using many fixtures. Generally, materials will be purchased at a plumbing supply store, although you will probably purchase your tools at your favorite hardware store; appliances are often purchased at appliance, furniture, or department stores. Try to purchase all your plumbing supplies and tools on one shopping trip to avoid interruptions while making an installation.

Tools

Tools for plumbing	$ _____
Tools for dismantling	$ _____
Tools for repair work	$ _____
SUBTOTAL	$ _____

Plumbing Supplies/Hardware

Pipe (in nominal size) with an allowance for errors	$ _____
Fittings, connectors, and adapters	$ _____
Materials for connecting pipe (pipe-joint compound, solder, solvent-cement)	$ _____
Pipe supports	$ _____
SUBTOTAL	$ _____

Fixtures, Devices, and Appliances

Plumbing fixtures (sinks, toilets, tubs, lavatories)	$ _____
Plumbing devices (such as filters, faucets, and valves)	$ _____
Plumbing appliances (such as a clothes washer or dishwasher)	$ _____
SUBTOTAL	$ _____

Repair Materials

Caulking Compound	$ _____
Wood putty	$ _____
Drywall	$ _____
Paint	$ _____
SUBTOTAL	$ _____
TOTAL	$ _____

GLOSSARY OF HOME PLUMBING TERMS

ABS. A rigid plastic pipe used for drainage. Acrylonitrile-butadiene-styrene.

Adapter fitting. A fitting that connects two pipes of different sizes or materials.

Aerator. A device snapped or screwed onto the end of a faucet. It mixes air with flowing water to minimize splashing.

Air chamber. A vertical length of capped pipe attached to a supply pipe near an outlet to prevent water hammer.

Backflow. A reversal of flow; can occur in the drainage or supply system.

Back-siphonage. The backflow of used or contaminated water into a potable supply pipe due to negative pressure in the pipe.

Ball cock. A valve inside a toilet tank that controls flow of water into the tank.

Branch. Any part of a plumbing system other than a riser, main, or stack.

Branch main. A water supply pipe that extends horizontally off a main or riser to convey water to branches or fixture groups.

Branch vent. A vent that connects two or more individual vents with a vent stack.

Cap. A solid cover used to close off the end of a pipe.

Cavitation. The formation of bubbles in a pipe caused by excessively high pressure.

Cleanout plug. A removable plug providing access to a drainpipe or a trap.

Closet bend. A drainpipe that joins the toilet bowl outlet with the drainpipe or soil stack.

Code. Legal requirements by which plumbing must be installed.

Compression fitting. A mechanical fitting used with copper or plastic pipe.

Coupling. A fitting used to connect two pipes in a straight run.

CPVC. A plastic pipe used for hot water lines. Chlorinated polyvinyl chloride.

Critical distance. The maximum horizontal distance allowed between a vent or soil stack and a fixture trap.

Diameter. The nominal inside pipe diameter as designated in the commercial sizing of pipes.

Diverter valve. A valve that changes the direction of water flow from one faucet or fixture to another.

Drain. A pipe which carries water-borne wastes in a building drainage system.

Dry well. See *seepage pit.*

DWV. Abbreviation for drain, waste and vent piping.

Effluent. The liquid that is discharged from a septic tank.

Elbow. Any fitting that connects two pipes to each other at an angle.

Escutcheon. A decorative piece that fits over a plumbing fixture, such as at a faucet body or a pipe coming out of a wall.

Female. Threading on a pipe, valve, or fitting that is internal.

Fitting. Any device that connects one pipe to another pipe or a fixture; allows for sizing or directional changes.

Fixture unit. A measurement of the load-producing effects on the plumbing system by different kinds of fixtures.

Fixture-unit flow rate. The total discharge or flow in gallons per minute of a fixture divided by 7.5 which provides the flow rate of that fixture.

Flow pressure. The pressure reading in a supply pipe near the faucet or outlet while it is open and flowing.

Hose bibb. A wall hydrant, sillcock, or similar faucet with a downward-angled threaded nozzle.

Hot water. Water at a temperature between 110° and 140° Fahrenheit.

House drain. The lowest piping that collects discharge from all other drainpipes in the house and conveys it outside the house.

Individual vent. A vent pipe installed for a single fixture; it connects with the vent system above or terminates individually outside the building.

Lavatory. A hand basin. Also, a bathroom.

Main. The principal pipe to which branches are connected.

Male. Threading on a pipe, valve, or fitting that is external.

Nipple. A short length of galvanized pipe that has external threads on both ends.

O-ring. A rubber ring used in some faucets to prevent leaks around a stem or base of a spout.

Packing. A soft material, such as graphite, pressed around a valve stem to prevent leaks.

Packing washer. A washer composed of packing material.

PB. Flexible plastic tubing used for hot and cold water supply lines. Polybutylene.

PE. Flexible plastic tubing used for cold water lines. Polyethylene.

Pipe sleeve. A clamp used to patch pipe leaks.

Pitch. See *slope.*

Plumbing appliance. An energized household appliance with plumbing connections. Examples: dishwasher, clothes washer, water heater.

Plumbing appurtenance. A device which is an adjunct to the basic system and demands no additional water supply or adds no additional discharge load to the system. Examples: water filter, relief valve, aerator.

Plumbing fixture. A device or receptor requiring a supply connection plus a discharge to the drainage system. Examples: toilet, lavatory, bathtub.

Plumbing system. Consists of the water supply and distribution pipes, plumbing fixtures, supports, and appurtenances; soil, waste, and vent pipes; sanitary drains and building sewers to an approved point of disposal.

Potable water. Water that is satisfactory for drinking.

PP. Rigid plastic pipe used for drainpipes and traps. Polypropylene.

Reducer. A fitting which enables pipes of different diameters to fit together.

Relief valve. A safety device that automatically releases water due to an excessive buildup of pressure and temperature; used on a water heater.

Riser. A water supply pipe which extends vertically one full story or more to convey water to branches or fixture groups.

Rough-in. The installation of all parts of the plumbing system which must be completed prior to the installation of fixtures. Includes DWV, supply piping, and built-in fixture supports.

Run. A vertical or horizontal series of pipes.

Sanitary cross. A drain fitting with two inlets; used in back-to-back installations.

Seat ring. A washer-like seal used in the construction of some faucets.

Secondary venting. Running an additional or second vent stack up and out the roof.

Seepage Pit. A pit into which uncontaminated water is drained.

Septic tank. A water-tight receptor receiving the discharge of a home's sanitary drainage system and designed to separate solids from liquid.

Siphoning. The creating of a vacuum in a pipe which acts to pull nearby water into it.

Slope. The fall of a line of pipe in reference to a horizontal plane. Expressed in a fraction of an inch per foot length of pipe.

Spacer. A short piece of copper or plastic pipe cut to size.

Stack. Any main vertical DWV line, including an offset, that extends one or more stories.

Standpipe. A special vertical drainpipe; often used for clothes washer hookups.

Stubout. A short drainpipe or supply pipe that sticks out of a wall or floor.

Sump. A tank or pit which receives sewage or liquid waste, located below the normal grade of the gravity system and which must be emptied by mechanical means.

Supports. Devices for supporting, hanging, and securing pipes, fixtures, and equipment. Also called *hangers* or *anchors.*

Sweating. Sweat soldering; connecting copper tubing with solder and a propane torch. Also, condensation on fixtures and pipes caused by warm air meeting a cold surface.

T fitting. A fitting into which a branch enters at a right angle.

Transition fitting. A fitting that joins two pipes made of different materials.

Trap. A fitting, either separate or built into a fixture, which provides a liquid seal to prevent the emission of sewer gases into the home.

Union. A fitting that joins two lengths of galvanized pipe and permits assembly and disassembly without taking an entire section apart.

Vacuum breaker. A device that prevents back-siphonage into a water line, typically through a threaded hose connection.

Valve seat. A ring inside a faucet body into which a washer or other piece fits, to stop the flow of water.

Vent stack. A vertical vent pipe installed to provide circulation of air to and from the drainage system and which extends through the roof.

Vent system. Piping installed to equalize pneumatic pressure in a drainage system.

Washer. A disc made of soft material that provides a seal against the flow of water.

Water hammer. The sound made by a banging pipe; caused by the sudden stoppage of water flow.

Water main. A water supply pipe for public use.

Y fitting. A fitting that has three outlets with the branch intersecting not at a right angle but at a slight angle as in the letter 'Y'.

INDEX

INDEX/CONT'D